T0215823

BestMasters

Springer awards „BestMasters" to the best master's theses which have been completed at renowned Universities in Germany, Austria, and Switzerland.

The studies received highest marks and were recommended for publication by supervisors. They address current issues from various fields of research in natural sciences, psychology, technology, and economics.

The series addresses practitioners as well as scientists and, in particular, offers guidance for early stage researchers.

More information about this series at http://www.springer.com/series/13198

Inga Lilge

Polymer Brush Films with Varied Grafting and Cross-Linking Density via SI-ATRP

Analysis of the Mechanical Properties by AFM

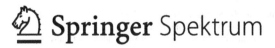 Springer Spektrum

Inga Lilge
Siegen, Germany

BestMasters
ISBN 978-3-658-19594-6   ISBN 978-3-658-19595-3   (eBook)
https://doi.org/10.1007/978-3-658-19595-3

Library of Congress Control Number: 2017952937

Springer Spektrum

Printed on acid-free paper

This Springer Spektrum imprint is published by Springer Nature
The registered company is Springer Fachmedien Wiesbaden GmbH
The registered company address is: Abraham-Lincoln-Str. 46, 65189 Wiesbaden, Germany

Acknowledgements

I would like to express my gratitude to all those who helped me out to finish this thesis.

First, I would like to thank Prof. Dr. Holger Schönherr for his constant support, guidance and help. I am especially thankful for the help of Dr. Davide Tranchida and his useful comments on the AFM experiments.

Moreover, I like to express my sincere thanks to all members of the Physical Chemistry I, for their valuable suggestions and kindness during all days of my thesis. Especially, I would like to thank my collegues for their valuable help for the experiments, endless discussions during coffee breaks and the presence on long days in the syntheses laboratory. At the same time, I express my sincere appreciation to Dipl. Ing. Gregor Schulte for the helpful discussions, the construction of the UV-LED setup and the fabrication of numerous gold substrates. And also to Dr. Lars Birlenbach for encouraging conversations to retry the fabrication of PAAm polymer brushes via PMP.

I give my warmest thanks to my parents for their constant support, motivation and encouragement especially during the last 8 months.

List of Content

List of Figures

List of Tables

List of Acronyms

σ	grafting density
δ	cantilever deflection
θ	contact angle
ε_λ	absorption coefficient
ν	Poisson ratio
AAm	acrylamide
AFM	atomic force microscopy
ATR	attenuated total reflection
bisAAm	N,N'-methylenebis(acrylamide)
c	concentration
CA	contact angle
CCD	charge-coupled device
CM	contact mode
d	distance between the chains
D_{sens}	deflection sensitivity
DTC	diethyldithiocarbamyl
E	Young´s modulus; elastic modulus
E_λ	reflection
F	applied load
f_1 and f_2	fractional areas of the components
FTIR	Fourier transform infrared
GPC	gel permeation chromatography
h	film height
I	intensity of the transmitted light
I_0	intensity of the incident light
IR	Infrared
k_n	cantilever elastic constant
l	path length
L	ligand
m	slope
MHDA	16-mercaptohexadecanoic acid
MMA	methyl methacrylate
MUBiB	ω-mercaptoundecyl bromoisobutyrate
ODT	1-octadecanethiol
OEG	oligo(ethylene glycol)
OEGMA	oligo(ethylene glycol) methyl methacrylate
p	penetration depth
PAAm	polyacrylamide

PDMS	poly(dimethylsiloxane)
PEG	poly(ethylene glycol)
PMP	photoiniferter-mediated polymerization
PSD	photo sensitive diode
Q	swelling ratio
RA	average roughness
RAFT	reversible addition-fragmentation transfer polymerization
R_g	radius of gyration
RMS	root mean square
R_{tip}	tip radius
S	standard deviation
SAM	self-assembled monolayer
SEM	scanning electron microscopy
SI	surface-initiated
SIP	surface-initiated polymerization
TM	tapping mode
TS	template-stripped
UV	ultra violet
z	piezo displacement in z-direction

Abstract

This thesis focuses on a systematic study of poly(acryl amide) (PAAm) brushes prepared by surface-initiated atom transfer radical polymerization (SI-ATRP). In addition to the analysis of the time dependence of the polymer brush growth, the conformation of the polymer brushes will be varied by grafting or cross-linking density. The grafting density is varied by the use of mixed self-assembled monolayers (SAMs) comprising of 16-mercaptohexadecanoic acid (MHDA) and ω-mercaptoundecyl bromoisobutyrate (MUBiB) to initiate the controlled radical polymerization from gold substrates. The cross-linking density is varied by an additional feed of a cross-linker, N,N'-methylenebis(acrylamide) (bisAAm), in the polymerization solution. The film thicknesses of polymer brushes in the dry state were obtained by ellipsometry and the surfaces were characterized by water contact angle measurements and FTIR spectroscopy.

It was found that mixed SAMs have a preferential adsorption of the thiol with the dummy functionalized head group (MHDA). SAMs composed of 45% initiator (MUBiB) on the surface have an initiator density, beyond which the polymer brush thickness no further increases. For lower values, the film thickness decreases to a minimum of 10 nm. Finally, the density of polymer brushes depends on the initiator density at the beginning of the polymerization. The polymer brush film thickness is also strongly influenced by the cross-linking density in an exponential manner, whereas the fraction of cross-linker in the polymerization solution reduces the resulting thickness of the polymer brushes.

Force curve measurements with atomic force microscopy (AFM) revealed that the variation of cross-linking density of polymer brushes influences the mechanical properties and swelling characteristics. In this case, the stiffness of the polymer brushes increases with increasing fraction of cross-linker and the swelling ratio decreases. These facts provide an interesting feature to control and tune the mechanical properties of surface grafted polymer brushes.

All in all, the tunable mechanical properties and reduced amounts of reactant molecules on the surface open a new way of varying surface interactions with cells, which are unknown to date and an interesting feature to investigate.

1 Motivation

The interest in techniques to produce thin chemically bound polymeric layers on solid surfaces has increased since 1990. Due to the chemical bonding of the polymer, chemical and environmental stability is enhanced over a wide range of conditions (pH, solvent, temperature).[1] In recent years polymer brushes have been widely used to design and modify surfaces with the focus on surface properties such as biocompatibility, dielectric property, wettability, corrosion resistance and friction. These surface modifications with polymers are especially important in printing, coating, food packaging, microelectronics, biomedical and aerospace applications.[2]

In particular it is widely known that surface coatings of poly(ethylene glycol) (PEG), oligo(ethylene glycol) (OEG) and their derivatives are used as anti-fouling materials, which resist protein adsorption, microbial and cell adhesion and have been studied extensively for biomedical applications.[3,4] However, PEG is susceptible to oxidative degradation, especially in the presence of oxygen and transition metal ions, thus their long-term applications are limited. According to this, PEG is non-toxic and shows no antigenicity or immunogenicity. All these factors expended together the basic knowledge of biocompatible surfaces.[5]

Acrylamide (AAm) is also known as an ultralow fouling surface coating, which highly resists protein absorption, controls cell adhesion and bacterial attachment. This polymer has already been commonly used in the form of a hydrogel for a variety of applications such as protein separation, drug release and water-oil separation due to its hydrophilic nature.[6,7] Polymer brushes of AAm are a fascinating class of materials because of the earlier mentioned unique monomer properties and the additional properties of a polymer brush. It is possible to fine-tune the properties, like thickness, grafting and cross-linking density of the polymer brush layers, which are key factors to influence their anti-fouling behavior.[6,8,9] This implicates the utilization of AAm-functionalized polymer brushes in this thesis, which are easily synthesized by surface-initiated polymerizations (SIP).[10]

An example for SIPs is the *photoiniferter-mediated polymerization (PMP)*, which has been employed extensively in the past decade to produce well defined block and graft copolymers. The ability to incorporate distinctly different chemistries within the same macromolecule has led researchers to pursue photoiniferters for grafting from polymeric surfaces. The photoiniferter molecules include a diethyl dithiocarbamate functionality, which contains bonds that cleave upon exposure to light and a silane functionality, which bounds to a silicon substrate.[11] PMP has been reported as a successful approach for the controlled grafting of a wide variety of monomers at ambient temperature and in aqueous media. The mild conditions required for this polymerization technique and the fast growth rates have made SI-

PMP attractive.[12] But all these photopolymers have an undesirable feature - the high toxicity of their components and their low environmental compatibility.[13]

Polymerizations of PAAm brushes on Silicon wafers have been performed with a silane-derivatized dithiocarbamate iniferter developed by de Boer et al.[2] It is already known, that silanes are very vulnerable to oxygen and water, which makes the fabrication of self-assembled monolayers (SAMs) and even structured SAMs for polymerization very difficult.[14] Matsuda et al. used a photomask during PMPs to restrict the incident UV light. With this technique, the formation of polymer is only allowed in the exposed area, directly from the liquid monomer.[15] The fabricated patterned brushes have a high lateral resolution, which is limited by optical phenomena.[16]

On the other hand, structured SAMs of thiolates exhibit a robust surface layer with a wide range of preparations, which is more stable to the surrounding environment and even the lateral resolution is less limited. The structured SAMs can be prepared by micro-contact printing (μCP), two-dimensional gradients or scanning probe microscopy techniques. These structured SAMs imply a specific end-functionalized moiety, which delivers a reactive center for a surface-initiated polymerization.[17] However these thiol SAMs still have several limitations. They contain only one single layer with limited robustness due to defects and the thiolates slightly tend to oxidize.[18] They also possess significant problems because of the thermally and UV-labile sulphur-gold bond, which is not reasonable when the advantages of thiol SAMs (high surface density and ease of formation) are combined with polymer brushes prepared by *atom transfer radical polymerizations (ATRP)*.[19] Then, the resulting film is thicker, more robust with a versatile architecture and chemistry.[10,20]

In ATRP, the direct initiation of polymer chains from a surface leads to high surface grafting densities (monomers can more easily diffuse toward the reactive center), whereas the grafting or selective adsorption of polymers is limited by steric and entropic forces. A route to control the density of surface-initiated polymer brushes is achieved by varying the concentration of the initiator molecules in SAMs.[2] The mixed SAMs of two alkanethiols, respectively one with and one without a terminal ATRP initiator group, allow a systematic control of the amount of initiator on the surface.[18] This will provide important information about the growth mechanism of surface-initiated polymer brushes as well as their density and morphology. Therefore, in this thesis the initiator density will be controlled quantitatively by using mixed-thiol SAMs on gold.

Huck et al. used mixed SAMs of undecanethiol and ω-mercaptoundecyl bromoisobutyrate for the controlled radical polymerization of methyl methacrylate (MMA).[2] Also Chilkoti et al. controlled the ATRP initiator surface density systematically by mixed SAMs on gold, which were prepared with undecanethiol and later on polymerized with oligo (ethylene glycol) methyl methacrylate (OEGMA).[18,5] The group of Genzer prepared different initiator densities by gradients generated with 1-trichlorosilyl-2-(m-p-chloromethylphenyl)ethane (CMPE) on silicon substrates and polymerized acrylamide (AAm) by ATRP. The resulting trend, shown in Figure 1.1, appeared to be consistent with the ones observed by Chilkoti.[21]

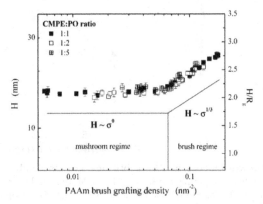

Figure 1.1: Wet thickness of PAAm brushes (H) as a function of the grafting density [Reprinted with permission from T. Wu, K. Efimenko, J. Genzer, *J. Am. Chem. Soc.* **2002**, *124*, (32), 9394-9395. Copyright 2014 American Chemical Society]

Another interesting feature is the variation of cross-linking density of surface-initiated polymer brushes, which is achieved by adding a changed concentration of cross-linker during polymerization. Recently, photoiniferter-mediated polymerizations have found application as a useful method for studying the structure and property evolution of highly cross-linked polymers, which was performed by Spencer et al.[12] In their study, PAAm brushes were grown by PMP from iniferter-functionalized silicon substrates with additional specific amounts of cross-linker, N,N-methylenebis-(acrylamide) (bisAAm). Later on, the swelling behavior and mechanical properties were linked to the cross-linking density of the PAAm brushes.[12]

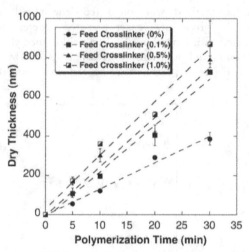

Figure 1.2: Growth kinetics of PAAm brushes and covalently cross-linked hydrogel brushes [Reprinted with permission from A. Li, E. M. Benetti, D. Tranchida, J. N. Clasohm, H. Schönherr, N. D. Spencer, *Macromolecules* **2011**, *44*, 5344. Copyright 2014 American Chemical Society]

It was found that bulk properties of these brushes are strongly influenced by the cross-linking of the polymer chains. The swelling ratios and the Young's moduli confirmed the formation of a network grafted from the surface, which is directly related to the amount of cross-linker in the monomer feed. [12]

In the beginning of this project, the preparation of PAAm brushes by PMP originally was carried out based on the procedure of the group of Spencer et al.[12] Several problems encountered during the synthesis of the iniferter and could not be solved in an appropriate time concerning the half year duration of a master thesis. In this case, an alternative synthesis strategy had to be selected. The preliminary results and the analysis of the problems are attached to this thesis (Chapter 8.1).

Later on, the positive aspects of surface patterning on gold and the variation of grafting or cross-linking density of AAm polymer brushes synthesized by SI-ATRP were combined. This will give an insight about the depending mechanical properties and the wettability, which can be linked to the conformation of the grafted PAAm brushes. The tunable mechanical properties and reduced amounts of reactant molecules on the surface open a new way of varying surface interactions with cells, which are unknown to date and an interesting feature to investigate.

The aim of this thesis is the investigation of synthesized PAAm brushes with an atomic force microscope (AFM), to determine, among other properties, the elastic modulus out of force curves. Hence, the polymerizations of pure, mixed and patterned initiator SAMs on gold surfaces were carried out in an aqueous AAm solution (with a cross-linker in the feed) at ambient temperature via SI-ATRP.

2 Introduction to Polymer Brushes

A lot of research has already been carried out focusing on polymer brushes, generated by surface initiated atom transfer radical polymerization (SI-ATRP). Polymer brushes are generally defined as thin layers of polymer chains end-grafted to a surface.[22,23] These polymer chains, which are terminally attached to a surface, have a distance between the chains (d), such that d is less than twice the radius of gyration (R_g) of the polymer. Due to chain crowding, the surface tethered macromolecules stretch away from the surface and they are in an entropic equilibrium when they exhibit an extended conformation. This conformation is defined as the equilibrium layer thickness (h), as demonstrated in Figure 2.1.[7–9]

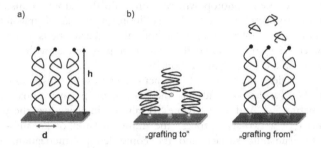

Figure 2.1: Schematic representation of a) polymer brushes and b) the "grafting to" and "grafting from" approach

Polymer brush layers are typically prepared through covalent attachment, which can be achieved through two available different experimental approaches, namely "grafting to" or "grafting from"-techniques, as shown in Figure 2.1. Both approaches have their unique characteristics.[24,25]

The "grafting to" approach implies the attachment of preformed polymer chains to a surface.[24] The end-functionalized polymer is reacted with reactive sites on a solid surface to form an anchored polymer layer. Once some preformed polymer chains are tethered to the surface, new polymer molecules have to diffuse through the existing attached polymer layer to reach surface active groups. The steric hindrance for surface attachment increases as the thickness of the polymer film increases. This technique is restricted due to the diffusion limited adsorption and steric hindrance of preformed polymers. As a result, low grafting densities are observed, because thick and dense polymer layers are difficult to achieve.[25,26,19] In contrast to the "grafting to" technique, the "grafting from" approach is known to be more suited to produce high grafting densities of chains and thicker films.[6] Due to the immobilization of initiating groups on the surface, the grafting density of chains is markedly increased. The polymer chains are grown *in situ* from a surface with a high density of initiating species and remain tethered to the substrate.[26,19,27] Because of these attractive properties, the "grafting from" approach has become the preferred option for the synthesis of polymer brushes with high grafting densities, whereas the surface immobilized initiator layer and subsequent *in situ* polymerization is referred to as surface initiated polymerization (SIP).[24]

2.1 Surface Initiated Polymerization (SIP)

In recent years, numerous experimental strategies have been developed for the preparation of polymer brushes by surface initiated polymerizations, such as photoiniferter-mediated photopolymerization (PMP), atom transfer radical polymerization (ATRP) and reversible addition-fragmentation transfer (RAFT) polymerization. These polymerization methods allow the synthesis of a wide variety of macromolecules with monomers that cannot be polymerized with the traditional cationic and anionic living polymerization.[23,28,29]

The covalent attachment of polymer chains to solid substrates by SIP is an effective method for the preparation of thick and stable polymer films with tailored surface properties.[29,30] And as the name SIP already implies, does the preparation usually consist of a two-step process the immobilization of an auto-associating group on the substrate surface and the controlled polymerization. The auto-associating molecule is modified with an initiator group.[2,26,31]

2.1.1 Photoiniferter-mediated Polymerization (PMP)

The photoiniferter-mediated polymerization explored by Otsu et al. in the early 1980s, is based on the formation of a reactive radical and a relatively stable counter radical, the iniferter (**ini**tiator-trans**fer**-**ter**minator agent).[32] The iniferter molecule includes a diethyldithiocarbamate functionality, which contains a bond, that cleaves upon exposure to light. One of the radicals resulting from the cleavage, the diethyldithiocarbamyl (DTC) radical, doesn't participate in the initiation, but reacts as a transfer agent and terminating species.[1]

In addition to the less reactive DTC moiety, the photoiniferter dissociates also into a highly reactive carbon radical, which acts as a controlled radical initiator without the addition of any other initiating species. Under correct conditions the DTC radical will cross-terminate with propagating macroradicals (initiated from the carbon radical). And the DTC groups are re-formed as the terminal groups on the polymer chain. Upon further exposure to light, the terminal iniferter species will cleave again and the re-initiation allows more monomer units to insert into the growing chain before another cross-termination occurs.[11] Usually UV labile C-C or

C-S bonds are used as DTC derivates.[33,34] Figure 2.2 shows a brief, idealized description of the overall mechanism of a photoiniferter-mediated polymerization.

$$R\text{-}DTC \xrightarrow{h\nu} R^* + DTC^* \qquad \text{Photo-Cleavage}$$

$$R^* + M \rightarrow RM^* \qquad \text{Initiation}$$

$$RM^* + nM \rightarrow RM_n{}^* \qquad \text{Propagation}$$

$$RM_n{}^* + DTC^* \rightarrow RM_n\text{-}DTC \qquad \text{Cross Termination}$$

$$RM_n\text{-}DTC \xrightarrow{h\nu} RM_n{}^* + DTC^* \qquad \text{Re-Initiation}$$

$$RM_n{}^* + xM \rightarrow RM_{n+x}{}^* \qquad \text{Propagation}$$

$$RM_{n+x}{}^* + DTC^* \rightarrow RM_{n+x}\text{-}DTC \qquad \text{Cross Termination}$$

Figure 2.2: Mechanism of the photoiniferter-mediated polymerization (PMP)

Propagation occurs by addition of monomers or double bonds to the active centers. The iniferter polymerization mechanism involves two termination pathways: first, the carbon-carbon radical combination and second, the carbon-DTC radical termination. When the viscosity of the polymerizing system is lower and thus, diffusion resistance to termination is low, path one is the significant termination mechanism. [1,13] If the photoiniferter molecules are chemically bound to a surface of a substrate, a polymer chain can be generated in the presence of monomer and during exposure to light. This enables the synthesis of thick polymer brush layers in a relatively short reaction time. Furthermore, the ability to reinitiate the polymerization promotes highly controlled thickness and chemistry of the grafted layer.[16]

Advantages of this polymerization technique include the facile control of the polymerization reaction by means of irradiation time and UV intensity, which leads to a linear increase of molecular weight (steady growth) with comparatively fast reaction kinetics compared to other methods. This method is compatible with a wide variety of monomers (acrylates, styrenes, acrylonitril and derivatives) and opens the possibility to form block copolymers by reinitiating the polymerization in a different monomer solution.[11] The reaction can be easily performed at room temperature or below, to avoid thermal polymerization and in aqueous media. Additionally, no sacrificial initiator in the monomer solution (limits the formation of free polymers in the bulk solution) required. The technique allows a precise control of the polymer molecular weight and usually yields to polymer brushes with low polydispersity.

2.1.2 Atom Transfer Radical Polymerization (ATRP)

In recent years, much attention has been paid to the use of surface initiated atom transfer radical polymerization.[23,28,31,35,36] Because of the ease of initiator formation and relatively mild polymerization conditions (at ambient temperature in aqueous media), it has been a powerful synthetic strategy to grow polymer brushes from various surfaces, such as gold.[37] In comparison to other controlled radical polymerization techniques, this polymerization is chemically extremely versatile and robust.[38]

ATRP is an effective and convenient method, which tolerates the use of many functional groups, and is easy to apply and results in highly uniform, dense polymer brushes with controlled thickness and composition on surfaces. [6,39-41] The thickness of the film can be adjusted by simply varying the polymerization conditions (i.e., time, monomer concentration, temperature).[35] The main principle of the ATRP reaction is a reversible redox activation of a dormant alkyl halide-terminated polymer chain end by a halogen transfer to a transition metal complex.[36]

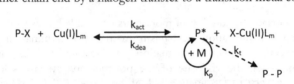

Figure 2.3: Mechanism of the atom transfer radical polymerization (ATRP)

The formal homolytic cleavage of the carbon-halogen bond, which results from this process, generates a free and active carbon-centered radical species at the polymer chain end. This activation step is based on a single electron transfer of the transition metal complex to the halogen atom, which leads to the oxidation of the transition metal complex. Then in a fast-reversible reaction the oxidized form of the catalyst reconverts the propagating radical chain end to the corresponding halogen capped dormant species.[36] In a typical ATRP reaction, CuCl or CuBr is generally used as a catalyst and bipyridines or multidentate amines as a ligand (L). The ligand must have a high complexation constant to compete with the polyacrylamid for copper, and it must allow fast redox between Cu(I) and Cu(II). Two mechanisms are possible for termination, first a radical transfer can occur from the intended surface bound radical to an AAm monomer in solution and second the Cu(II) species in the solution can act as the catalyst for the extraction of HX in the atom-transfer living polymerization.[42,43]

Significant effort has gone into achieving controlled SI-ATRP. In general, this can be achieved either by adding sacrificial initiator or by adding a Cu(II) salt as a deactivator. In the former case of the sacrificial initiator, a persistent radical is created artificially, because the active amount of conversion and living chains on the surface is too small to establish the standard equilibrium present in bulk ATRP. The disadvantage is the formation of bulk polymers that are difficult to remove.[44] The addition of Cu(II) can be a good alternative, but adding very small amounts of both Cu(I) and Cu(II) salts in a precise ratio is difficult, because of the oxygen sensitivity of Cu(I). The quenching of the polymerization with Cu(II) species helps to preserve initiation points.[39]

Already a lot of research has been carried out focusing on polymer brushes generated by SI-ATRP. There is no universal recipe and many factors, including the choice of solvents, ligands, monomers and reaction temperature, will affect the polymerization.[45] These parameters can influence the performance of the polymerization, but also offer in the same time a wide range of possibilities to fine tune the reaction.[5] Overall requires the synthesis of polymer brushes by SI-ATRP self-assembled monolayers (SAMs) composed of an active species (the initiator) to graft polymer brushes from a surface.

2.2   Self-Assembled Monolayer (SAM)

A SAM is defined as a monomolecular film of a surfactant formed spontaneously on a substrate upon exposure to a surfactant solution.[46] These organic assemblies are formed by the adsorption of molecular constituents having a chemical functionality or headgroup, with special affinity for a substrate.[47] The main chain, also called tail, forms a highly ordered structure oriented away from the substrate, organized by van der Waals interaction and a terminal group, which determines the surface chemistry.[48,49] The main driving forces of the formation of these films are the specific interactions between the reactive head group and the surface of the substrate. Based on these interactions, the SAMs are stable films with a remarkable robustness.[46]

The most molecular monolayers are thiol based adsorbates on gold, which are employed in order to synthesize functional surfaces, to serve as initiator for polymerizations, focusing on covalently attaching a monolayer with high grafting densities on the surface.[10] On surfaces such as glasses, silicon wafers, quartz and mica, an alkoxysilane group is converted into a stable poly(siloxane) layer by coupling with the free hydroxyl group.[2,49] But SAMs have still several limitations. Due to the self-assembling nature of their formation, it is practically impossible to obtain large-area defect-free monolayers. And the monolayers, which are mechanically and chemically fragile, are only several nm thick. In the other case, SAMs can be used to introduce almost any functional group to the surface. Therefore, initiator terminated SAMs provide a conceptionally simple route to grow polymer brushes as robust, functional and switchable surfaces.[50]

The surface wettability, roughness and chemistry are important parameter, which mainly depend on the introduced chemical functionality of the tails of the thiol molecule. As a result, SAMs provide a simple and convenient way to change surface parameters by altering the terminal groups.[50] If a sample is exposed to a solution containing two different kinds of thiols, a co-adsorption will take place and a binary or mixed monolayer (mixed SAM) will be formed.[51] Generally, the composition of the surface of these mixed SAMs after self-assembly is not identical with the one in the binary solution.[52] In this case, the mixed SAMs could have a non-ideal behavior between the surface and solution compositions.[20,53,54] Another way to introduce two different kinds of thiols onto a gold surface, can be achieved by micro-contact printing (μ-CP).

2.3   SAMs by Micro-Contact Printing (μ-CP)

In recent years, the fabrication of patterned materials on surfaces at ever smaller length scales became more and more interesting, when electrical circuit designs rapidly increased in size and complexity.[55] Photolithography, which is a technology used to generate the initial patterns in electronic circuits, has been extended to pattern other materials. This technology can also be applied to generate patterns of different surface chemistries and is used to control the spatial distribution of other materials. In particular, μ-contact printing is a technique with rapid preparation of substrates as well as patterning of a wide range of materials.[56]

μ-Contact printing is a flexible new technique that forms patterned SAMs with regions terminated with different chemical functionalities, and thus different chemical and physical properties, in patterns with μm dimensions.[55,57] It has proven to be a useful technique and particularly valuable in the patterning of biological materials.[55] In this method, first described by Whitesides group in 1993, an elastomeric "stamp" is used to transfer an alkanethiol "ink" onto a variety of substrates, for example gold, silver and silicon dioxide, by conformal contact between the stamp and the surface.[58,59] Patterns of binary SAMs, patterned SAMs terminated with one chemical functionality and a background of another, are formed by contact printing one type of molecule and backfilling with another.[59]

As already mentioned, photolithography is a versatile and precise technique, that routinely generates submicron features. Thereupon, the surface of a silicon wafer (A in Figure 2.4) is coated with a thin and uniform layer of a photoresist, an organic polymer which is sensitive to ultraviolet light (B in Figure 2.4). Then the photoresist is exposed to light through a metal photomask (C in Figure 2.4). The light passes through the mask, generating an area-selective polymerization (degradation) of the photoresist according to the pattern of the mask. Afterwards, the uncured polymer is removed (D in Figure 2.4), thus the master is completed and ready for the preparation of several elastomeric stamps. A poly(dimethylsiloxane) (PDMS) elastomer is typically used as a stamp to transfer the pattern from the master to the substrate. For the stamp preparation, a liquid vinyl-terminated prepolymer and the curing agent are mixed and the mixture is purred onto the patterned template. After curing the PDMS at 60°C, a solid elastomeric polymer is formed. Finally, the PDMS is peeled off, cut in proper size and used for the actual μ-contact printing.[55,56]

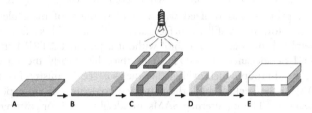

Figure 2.4: Schematic illustration of the stamp preparation for μ-contact printing by photolithography

Afterwards, the stamp will be loaded with the material that is to be printed, whereas the binding of the molecules to the new surface must be more energetically favourable, than staying on the stamp. According to the highly hydrophobic properties of PDMS only apolar materials can be used as ink solutions. For patterned surface properties, the most common ink solution that has been widely used is a thiol solution. Thiol solutions are SAM forming molecules, which allow the formation of a stable condition of molecules on the surface.

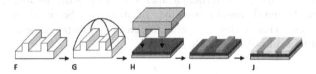

Figure 2.5: Schematic illustration of the actual μ-contact printing

The inking of the stamp is either achieved by immersing the stamp in the ink solution or placing a droplet of the ink solution on the patterned side of the stamp (G in Figure 2.5). Afterwards, the stamp surface is dried by solvent evaporation in a nitrogen gas flow.[59,56] Then, the inked stamp is brought into contact with the gold surface (H in Figure 2.5) and the thiol is transferred to the surface (I in Figure 2.5). During this process, the thiol is selectively transferred to the surface according to the pattern of the stamp, as only the areas with the protrusions of the PDMS stamp are able to contact the gold surface.[56] After some time the stamp was separated carefully from the gold surface. For the SI-ATRP reaction prints a bifunctional

33

initiator that contained a thiol at one end and the ATRP initiator at the other end (a bromoisobutyrate moiety) was used.

For the use of patterned substrates for polymerization by ATRP, the patterned surface is backfilled with a different thiol, to allow the formation of a stable condition of a SAM on the whole gold surface (J in Figure 2.5). When the μ-contact printing was realized with a thiol solution of molecules with terminal ATRP initiator, the unfilled surface areas of the gold surface are filled by a self-assembled monolayer of a thiol without a terminal ATRP initiator group, for example 1-octadecanethiol (ODT). This thiol has nearly the same length but a different head group which exhibits another surface chemistry. For the preparation of a SAM of the backfill thiol, the patterned gold-surfaces are exposed to a 1 mM thiol solution.[60] These patterned SAMs are ideally suited for grafting polymers onto surfaces because of the very high density of the functional groups, small number of defects and their well-defined structure. In addition, a systematic control of the spatial growth of polymer brushes on the surface is possible.

It has to be taken into account that undesired pathways for transport of thiols onto non-printed parts of the sample may exist. The transport of thiols through the stamp depends on the molecular weight of the thiol ink. Transport through the ambient air is proportional to the vapor pressure of the ink and vapor pressure decreases with increasing molecular weight.[61] A stamp deformation during stamp removal from the template and during the contacting of the substrate also has to be taken into account, which limits the resolution of the patterning.[56] But still this form of soft lithography technique always produces sharp boundaries between the distinct chemical regions on the substrate to be used for SI-ATRP polymerizations. This feature is useful for creating substrates with well-defined chemical patterns of polymer brushes.[35]

2.4   Variation of Polymer Brushes

Polymer brushes are a very attractive tool for tailoring interfacial properties and SI-ATRP is a very suitable synthesis strategy to grow these polymer brushes from a surface because it allows a good control over the brush thickness and grafting density.[41,62] In this case, the polymer layer thickness can be easily and precisely controlled by varying the polymerization time.[60,63]

2.4.1   Grafting Density

The variation of the grafting density is based on the modification of the substrate, more precisely on the resulting surface chemistry of a mixed SAM.[52] This approach focuses on covalently grafting thiol molecules onto the surface containing a defined amount of initiator moieties, whereas these macroinitiators are large compound molecules composed of both initiating and anchoring moieties.[7] The results of previous projects have shown, that the thickness of polymer brushes is linked to the initiator density on the gold substrate, which can be associated to the final polymer grafting density.[62,64] This aspect enables a new way to tune the thickness and the resulting properties of polymer brushes.

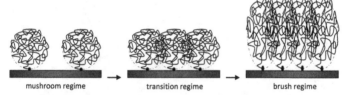

mushroom regime   →   transition regime   →   brush regime

Figure 2.6: Model of the varied grafting density of polymer brushes

If the graft density of polymer brushes is high, the polymer chains tend to be strongly stretched along the direction perpendicular to the substrate (brush regime, Figure 2.6). In the case of low densities, the "mushroom" chain conformation similar to that of the free chains will appear for the tethered chains.[65,66] With increasing grafting density, the chains are more and more forced to a brush morphology as a result of a balance between an entropic effect, caused by higher orientation of the chains, and an energetic effect due to minimization of the repulsive interactions.[24,45] Also steric factors can inhibit the polymer chains growing from every single initiator molecule in the SAM. The surface average cross sectional area of an initiator molecule (20 Å^2) is much smaller than the area of a polymer chain (~180-200 Å^2). In this case, only one out of 10 initiator molecules bound to the surface is expected to initiate a polymer chain.

2.4.2 Cross-linking Density

The variation of cross-linking density is based on the fractional feed of a cross-linker in the polymerization solution, which usually results in an insoluble three-dimensional polymeric gel.[29] Polymer gel brushes were previously prepared either in situ, by surface initiated polymerization in presence of a cross-linker, or ex situ by postmodification of presynthesiszed polymer brushes. [12,29, 67] The surface grafted polymer networks represent a novel class of thin hydrogel films (covalently cross-linked hydrogel brushes), whereas the ability to introduce and modulate the cross-linking density within surface grafted polymer brushes is expected to influence their mechanical and chemical stability, permeability and swelling characteristics. The change of lateral forces upon cross-linking depends on the dissociation kinetics of the cross-linkers, which therefore provide a molecular handle to control the mechanical properties of surface grafted polymer brushes.[67]

3 Characterization Methods

The characterization of the modified gold surfaces was performed by various methods, as light microscopy, ellipsometry, scanning electron microscopy (SEM), contact angle (CA) measurements, fourier transform infrared (FTIR) spectroscopy and atomic force microscopy (AFM). The last three characterization methods will be discussed in more detail by introducing the equations for the following analysis.

3.1 Contact Angle (CA) Measurements

Contact angle measurements are used to demonstrate the relationship between the properties and the chemistry of a surface. The ability of a fluid to cover a surface, the so-called wettability, varies with the perfection of the monolayer and with the degree of order. The wettability depends also on the polarity of the monolayer surface functional groups. In this case, the contact angle measurement is a method for surface analysis related to surface energy and tension. When a droplet of liquid rests on the surface of a solid, the shape of the droplet is determined by the balance of the interfacial liquid/vapor/solid forces.

The CA describes the shape of a liquid droplet lying on a solid surface (sessil drop). The contact angle is defined as the angle between the solid surface and the line obtained as tangent to the bubble profile at the angle between the solid surface and the line obtained as tangent to the bubble profile at the solid surface, on the side of the droplet, see Figure 4.1. The contact angles can be defined on ideal surfaces with the Young's equation (Equation (3.1)). In this case the interfacial tensions on ideal surfaces γ_{GL} (Gas-Liquid), γ_{GS} (Gas-Solid) and γ_{SL} (Solid-Liquid) result in an equilibrium angle θ, shown in the following equation:[68]

$$\gamma_{GL}\cos\theta = \gamma_{GS} - \gamma_{SL} \tag{3.1}$$

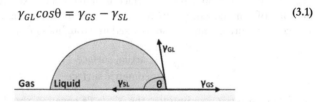

Figure 3.1: Equilibrium contact angle with the assigned notations of the Young's equation

A droplet of liquid will be dispensed onto the substrate surface and a Charge-coupled Device (CCD) camera captures the profile of the droplet on the computer screen. After a baseline is set, which identifies the solid surface, the software calculates the tangent to the droplet shape and the contact angle. On real surfaces, when the surface roughness has a significant influence, the so-called static contact angle is measured. In this case, the sample is moved upwards to pick the drop and immediately a picture of the drop is taken for off-line analysis. The dynamic contact angle can be defined at a drop rolling on a leaning surface. As an alternative, the contact angle can also be measured on a horizontal surface by increasing and decreasing the volume of the droplet. In this way, the advancing and receding contact angles are measured. They are summarized in the dynamic contact angle, which correlates better with the electrical properties of the surface and provides more information of the surface wettability.

The analysis of contact angles of mixed SAMs is based on the Cassie and the Israelachvili equation. With the Cassie equation, heterogeneous surfaces including impurities or polycrystallinity over small length scales can be modeled. It is presumed, that the surface is composed of well-separated and distinct patches or domains. The Cassie equation can be formulated as follows: [68,69]

$$cos\theta = f_1 cos\theta_1 + f_2 cos\theta_2 \quad (f_1 + f_2 = 1) \tag{3.2}$$

(f1 and f2: fractional areas of the components, θ: contact angle)

When the surface of the binary system cannot be represented by discrete patches, it is rather having a mixed homogeneous surface. The Israelachvili and Gee equation can be formulated focusing on the polarization and dipole moment:[70]

$$(1 + cos\theta)^2 = f_1(1 + cos\theta_1)^2 + f_2(1 + cos\theta_2)^2 \tag{3.3}$$

Another important feature, which has an influence on the resulting contact angle of a surface is the surface roughness, which was introduced by Wenzel. Within a measured unit area on a rough surface is actually more surface and thus a greater intensity of surface energy than in the same measured unit area on a smooth surface. This surface ratio is summarized in a roughness factor:

$$R = \frac{\text{actual surface}}{\text{geometric surface}} \tag{3.4}$$

According to these assumptions, the Young's equation should be modified by the multiplication of the observed contact angle on a smooth surface by the roughness factor which is displayed as follows:

$$cos(\theta_{rough}) = Rcos(\theta_{smooth}) \tag{3.5}$$

The dependency of the contact angle on the surface roughness, whereas the roughness results in increased contact angles for smooth contact angles that are less than 90° and decreased ones for smooth contact angles greater than 90°.[71,72]

3.2 Fourier Transform Infrared (FTIR) Spectroscopy

Infrared spectroscopy is mostly used to determine the chemical functional groups within a sample, because different functional groups absorb characteristic frequencies of IR radiation. The IR radiation may be passed through a sample or may be reflected from a sample and a whole spectrum can be measured at all wavenumbers in the range of 400-4000 cm^{-2}. The spectrum represents the molecular absorption and will create a molecular fingerprint of the sample. The absorption peaks correspond to the frequencies of vibration between the bonds of the atoms inside the material. In addition, the size of the peaks in the spectrum is a direct indication of the amount of the compound. In this case the areas of certain peaks can be calculated and interpreted using Lambert Beer's law.

$$E_\lambda = -\lg\left(\frac{I}{I_0}\right) = \varepsilon_\lambda * c * l \qquad (3.6)$$

(E_λ=reflection; I_0=intensity of the incident light; I= intensity of the transmitted light; ε_λ= absorption coefficient; c= concentration; l= path length)

In this work, mainly reflectance spectra were recorded, which basically involves a mirrorlike reflection and produces reflection measurements for reflective materials (such as gold). A reflection-absorption spectrum of a surface film will be obtained. This way, thin surface films in between the range of nano- and micrometers can be routinely examined with an angle of incidence (grazing angle) of 80°, respectively. Due to the great incident angles, the radiation parallel polarized to the plane of incidence starts to interact with the surface film. The s-polarized components have no impact on the measurement. Thus, only the vibrations perpendicular to the surface will be detected. Due to an expansion of the incident angle, the sensitivity can be increased as also the corresponding absorption.

3.3 Atomic Force Microscopy (AFM)

An AFM opens a way to investigate the morphology of surfaces of films in a nanometer scale range as well as their physical properties. The most important part of an AFM is an elastic cantilever, which is attached to a piezoelectric crystal. At the end of the cantilever is a very sharp tip, to collect high resolution images or to indent the samples to measure mechanical properties.

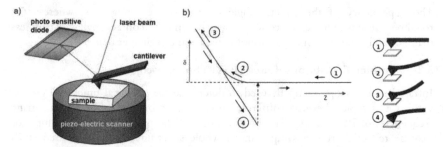

Figure 3.2: Scheme of a) an atomic force microscope (AFM) and b) Scheme of a typical cantilever deflection vs. piezo position curve.

The resonant frequency of the cantilever has to be identified before every experiment. Using this frequency, the piezocrystal, where the cantilever is attached, oscillates. In tapping mode, the tip steps with constant oscillation towards the surface. In case of interactions of the cantilever with the sample, the oscillation decreases its amplitude and the laser beam changes its position along the optical detector, a photo sensitive diode (PSD). The PSD is connected to the computer and a feedback control. Simultaneously an image is generated, based on the signal of the PSD, in the sense that the piezo, with the attached sample, is moved up and down to keep the tip oscillation constant, and therefore an AFM (topography) image is a plot of piezo movements.

On the other hand, an AFM can also be used to perform nanoindentation measurements via force curves, which are obtained by recording the applied load on the tip with the corresponding penetration depth. During the nanoindentation, the threshold for penetration depths and the tip rate towards the sample are controlled by the software. Mainly the mechanical properties of soft matter as biological materials and polymers are recorded. In the case of polymers, the Young's modulus of samples can be studied. Before the previously obtained raw-force curves may be converted into a plot of applied load vs. penetration depth, several calibrations and preliminary determinations are essential. First the cantilever elastic constant, which is used to estimate the instantaneous applied load (F) from the experimentally available cantilever deflection (δ) in the following context.

$$F = k_n \delta$$

(F: applied load; k_n: cantilever elastic constant; δ: cantilever deflection) (3.7)

Secondly, the deflection sensitivity (D_{sens}), which is an important parameter, needs to be calibrated. Because the photodiode is used to monitor the cantilever deflection by an optical lever arm system, for which the voltage has to be calibrated so that it can be converted into cantilever deflection. Therefore, a hard material is used, that could not be indented by the tip, so that the cantilever deflection equals the piezo displacement. In this context, the deflection sensitivity is a conversion factor to translate the voltage into cantilever deflection, which is obtained from the slope of the plot of the voltage against the piezo displacement. It has to be taken into account that the calibration can vary strongly with the set-up and alignment of the cantilever.

After the preliminary determinations and calibrations are performed, the obtained raw-force curves are plotted into diagrams with the applied loads on the y-axis against the penetration depth on the x-axis. The penetration depth is calculated following equation:[74]

$$p = z - \delta$$
(p: penetration depth; z: piezo displacement in z-direction; δ: cantilever deflection)   (3.8)

In this work the elastic modulus for the given values of penetration depth and applied loads will be calculated using Sneddon's purely elastic contact mechanics model. By contrast to former theories, the main aspect of Sneddon's theory is affected by the model of the real shape of the tip in nanometer length scale. The others worked with a really sharp defined tip, like ideal pointed cones. The tip has a paraboloidal section, which means that the theory by Sneddon is still an approximation: [75]

$$F = Ap^{1.5}$$
(3.9)
(A: prefactor; p: penetration depth)

This equation shows that there is no proportional relationship between the applied load and the penetration depth. The exponent was chosen as 1.5, but in general it is a function of the investigated material and the geometry of the tip. In Sneddon's theory, the elastic contact between the thin film and the tip is included. The prefactor A contains the value of Young´s Modulus, shown in the next equation.[74]

$$F = \frac{4E}{3(1 - v^2)} 2\sqrt{R_{tip}} p^{3/2}$$
(3.10)
(E: Young´s Modulus; F: Applied load; v: Poisson ratio; R_{tip}: tip radius)

4 Experimental Part

4.1 Materials

Substrate:
> **Microscope slides:** purchased from Thermo scientific, Menzel-Gläser (ISO 8037/1; 76x26 mm)
> **Silicon slides:** Silicon (111) wafers (Okmetric wafers, thickness 525+/-25 μm)
> **HELLMANEX®II:** purchased from Hellma
> **Isopropanol:** purchased from J.T. Baker
> **Epoxy glue:** EPO-TEK 377 purchased from Polytec PT GmbH Polymere Technologien

Surface modification:
> **Gold:** purchased from Allgemeine Gold- und Silberscheideanstalt AG (Pforzheim); 99.99% (granules)

Chromium: purchased from Chempur; 99.998% (pieces: 1-6mm)

Cleaning of the wafers:

Piranha solution (1:3 (v/v) 30% H_2O_2 and concentrated H_2SO_4)

Hydrogen peroxide (H_2O_2): purchased from Roth; 30%

Sulfuric acid (H_2SO_4): purchased from Riedel de Haën, 95-97 % extra pure

Ethanol: purchased from Fischer Scientific, 97%, denatured

Chloroform: purchased from Roth; ≥99%, p. a.

PDMS preparation:
> **curing agent:** Silicone Elastomer 184, purchased from Sylgard
> **silicon polymer base:** Silicone Elastomer 184, purchased from Sylgard

Reaction solution:
> **Cu(I)Br:** synthesized in-house
> **PMDETA:** 1,1,4,7,7, pentamethyldiethylene triamine: purchased from Sigma Aldrich; ≥99%
> **Methanol:** purchased from J.T. Baker

FTIR reference:
> **Deuterated thiol:** synthesized in-house by Dr. E. Sperotto

4.1.1 Initiator

ω-Mercaptoundecyl bromoisobutyrate (MUBiB) was synthesized according to Jones et al. and is the initiator functionalized thiol, used to perform the SI-ATRP (Figure 4.1a).[64] The bromine based initiator was chosen to ensure a fast initiation, due to an easy cleavage of the bromide-carbon bond.[6] 16-mercaptohexadecanoic acid (MHDA) and 1-octadecanethiol (ODT) were obtained from Aldrich and used without further purification (Figure 4.1b and c). These compounds are so called "dummy" initiators and have a similar structure, but a different wettability than

MUBiB. MHDA SAMs are hydrophilic and have a static contact angle of around 10-15°, ODT SAMs are hydrophobic with a contact angle of 110°.[76,77] In this work ODT was chosen for μ-contact patterns, due to its high surface affinity. The molecular weight of ODT (286 g/mol) has an effect on the transport, because it limits the diffusion of the ink during printing. This results in a monolayer pattern, which is representative of the patterns of the stamp.[61] MHDA was used in this work for the preparation of mixed SAMs on the basis of an approximately similar surface affinity as the initiator thiol.

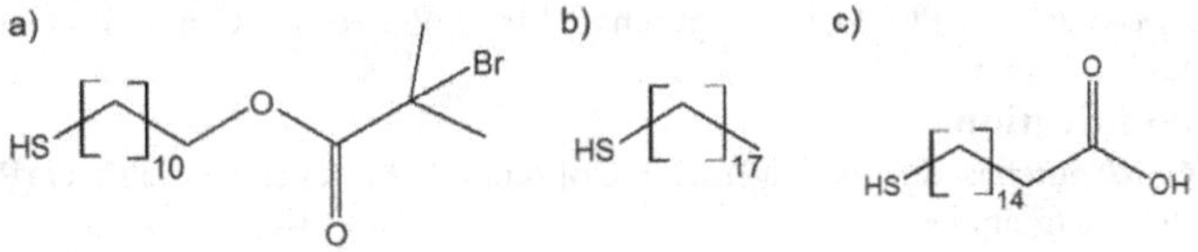

Figure 4.1: Molecular structure of the thiol molecules a) MUBiB [ω-Mercaptoundecyl bromoisobutyrate], b) ODT [1-octadecanethiol], c) MHDA [16-mercaptohexa decanoic acid]

4.1.2 Polymers

In this study a monomer, AAm [acrylamide] (99.9%) purchased from Sigma Aldrich, was used as received for the polymer brush synthesis (Figure 4.2a). For covalently cross-linked polymer brushes a specific amount of a cross-linker, bisAAm [N,N'-methylenebis(acrylamide)] was added. These polymers have different amounts of active centers in the molecule (Figure 4.2b).

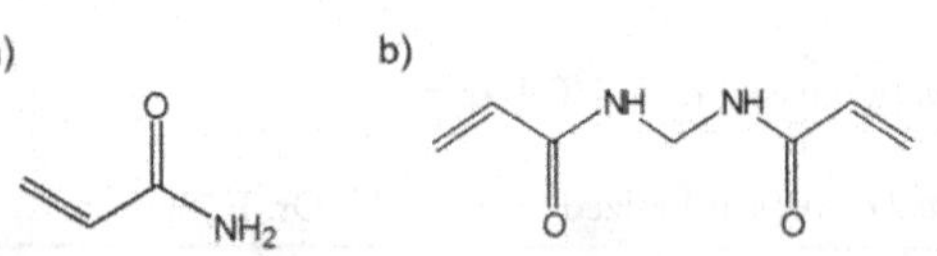

Figure 4.2: Molecular structure of the monomer a) AAm [acrylamide] and b) bisAAm [N,N'-methylenebis(acrylamide)]

4.2   Preparation

4.2.1   Preparation of Gold Substrates

In the beginning the glass substrates had to be cleaned before the gold evaporation process, to be able to obtain a homogeneous gold film on the surface. In addition, to avoid any scratches on the surfaces, the glass substrates with a diameter of 2.0 cm, were handled very carefully. In this case, the glass substrates were hold with tweezers only at the edges.

The glass substrates were cleaned with HELLMANEX®II at 2.0%. The substrates were kept in a beaker, filled with the HELLMANEX®II solution, placed in an ultrasonic bath (20°C; 15 min) filled with water. Afterwards, they were thoroughly rinsed with Milli-Q water and kept in a beaker of high purity water, in an ultrasonic bath for 5min. Subsequently the glass substrates were dried with a displacement drying method in an ultra-clean fume hood. An anhydrous fluid, isopropanol, was used to displace the water from the surface with a "vapor dry" technique. The cold glass substrate was immersed in the vapor above heated isopropanol. On the cold surface the alcohol vapor condensed and removed the water and other contaminants. Afterwards, the surface becomes hot and then the substrate is removed to make the surface rapidly dry.

The samples were directly placed into the sample holder of a gold sputtering machine (Edwards; E306A: high vacuum - vapor deposition with oil diffusion pump [YOM 1980] and electron beam vaporizer [EB3]). The substrates were primed with 10 nm Cr and coated with 120 nm Au (0.1 nm /sec; pressure around $8*10^{-6}$ mbar, T = 50°C). In order to obtain appropriate sizes for further functionalization, polymerization and analysis the gold substrates were cut in half with a diamond glass cutter. Until usage, these pieces were kept in a Petri dish sealed with parafilm.

4.2.1.1   Template-Stripped (TS) Gold Substrates

For the first step, silicon wafer were cut into 1 x 1 cm pieces with a diamond glass cutter. Then, the wafers were first dipped and washed in chloroform and additionally in ethanol. Afterwards, they were rinsed with Milli-Q water and then dipped for 2 minutes into the Piranha-Solution (1:3 (v/v) 30% H_2O_2 and concentrated H_2SO_4). After a second rinsing with Milli-Q water, the silicon wafers were dried with a clean and dry argon gas flow. The cleaned substrates were directly placed into the vacuum chamber of a thermal evaporation machine (MED 010, Blazers Union) and the pressure was allowed to decrease over night down to 10^{-6} mbar.[78]

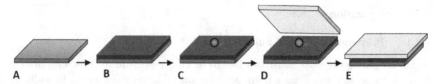

Figure 4.3: Schematic illustration of the preparation of template-stripped gold

The next day, amber colored glue (EPO-tek 377, epoxy glue) was prepared by mixing equal amounts (by weight) of both components, resin and hardener, together. Afterwards, the glue was stored under vacuum for two hours, to remove any bubbles inside the mixture. In the meantime, gold was evaporated on the silicon wafers with roughly 100 nm gold (Figure 4.3B). The evaporation rate, which was less than 0.2 Å/s and film thickness were monitored and controlled by a quartz crystal thickness controller. Meanwhile, the glass substrates where cleaned by the similar piranha cleaning step as the silicon wafers. Afterwards, the thermal evaporation machine was ventilated and the freshly gold coated silicon wafers could be glued to the glass substrates. Therefore, one tiny drop of glue was placed in the middle of the gold surface of the silicon wafer and the glass substrate was attached to the surface by sliding the piece onto the silicon wafer and in a preheated oven for 2 hours at 120-130 °C (Figure 4.3C and D).[79]

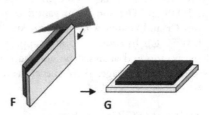

Figure 4.4: Schematic illustration of the stripping of a Si-Au-epoxy glue-glass "sandwich"

The resulting Si-Au-epoxy glue-glass "sandwiches" could be stored as stripping precursors without detectable loss of quality. Prior to use, the sandwiches were stripped from each other by applying a gentle force with a scalpel to the boarder of the two different wafers, as illustrated in Figure 4.4.[80,81,82] These template-stripped gold substrates do not require any cleaning step prior to use, for any functionalization. The gold substrates, obtained by thermal vapor deposition, had to be cleaned with piranha solution immediately before further usage in order to obtain a clean gold surface. The gold substrates were first dipped and washed in chloroform and in ethanol. Afterwards, they were rinsed with high purity water. The water was removed by shaking and then the substrates were placed for 3 minutes in the Piranha-Solution. After a second rinsing with high purity water and with ethanol the substrate, for SAMs by μ-CP, was dried with a clean and dry argon gas flow. For pure and *mixed SAMs*, the substrates were also rinsed with high purity water and ethanol after piranha treatment, but in this case the gold surface was capped with a drop of ethanol and directly transferred into the solution.

4.2.2 SAMs

The gold substrates were immersed in 5 mL alkanethiol solution (1 mM) in a snap-on lid glass. These solutions were prepared and could be stored in the fridge for a maximum of one month. Finally, the snap-on lid glasses with the gold substrates were flushed with argon and kept closed in a place with ambient temperature overnight.[49]

Mixed SAMs with different initiator concentrations on the surface were synthesized by varying the concentration of the dummy-initiator in the alkanethiol solution.[52] Therefore, the alkanethiol solution was prepared with a specific amount of the dummy-initiator MHDA ($W_{MUBiB}: W_{MHDA} = 0, 0.1, 0.2, \ldots 1.0$).

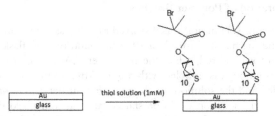

Figure 4.5: Scheme of the monolayer formation of the initiator on a gold substrate

4.2.3 Micro-Contact Printing (μ-CP)

For the fabrication of PDMS stamps 1.2 g of curing agent (10%) was filled up to 12 g of silicon polymer base (90%). Afterwards, a proper mixing of the PDMS components, which is essential for a good curing. Most of the trapped bubbles from mixing will eventually rise to the top of the liquid where they may be broken by blowing across the surface. A vacuum pump may help to remove the bubbles more quickly. Then, a silicon master is placed in a glass petri dish with a piece of paper in between and the metal tubing is placed around it. The uncured PDMS mix is slowly poured over the master, starting in the middle and allowing the PDMS to spread out evenly. After a thickness of roughly 1 cm is reached, the pouring can be stopped gently. The assembly is left at room temperature for a few minutes, so that incorporated bubbles during pouring can rise out of the PDMS. Afterwards, the whole assembly is placed in an oven, which is set to 80 °C for 60 min for the PDMS to cure. The mold is removed from the oven and allowed to cool for a few minutes until it is safe to handle. First, the metal tubing is removed and then the PDMS can be peeled from the master. A single-edge razor blade is used to cut out around the face of the stamp. The stamp can now be used for μ-CP.

Therefore, several drops of the alkanethiol solution were used to cover the surface of the stamps entirely. The solution is allowed to sit on top of the pattern of the PDMS stamp for about a minute. The excess is removed with a clean and dry nitrogen gas flow. For the actual printing step the dry stamp was turned face down and brought into contact with the freshly cleaned gold substrate by applying a gentle pressure across the entire stamp surface. For the backfill of the patterned gold surface, these patterned substrates were transferred as described before into a 5 mL MUBiB solution (1 mM) in a snap-on lid glass. Afterwards the snap-on lid glasses with the gold substrates were flushed with argon and kept closed in a place with ambient temperature for some time (~5 minutes). Finally, the SAMs were thoroughly rinsed with ethanol and high purity water to remove physisorbed molecules.

4.2.4   Preparation of Polymer Brushes

PMDETA (0.28 mL, 1.34 mmol) was dissolved in a degassed mixture of high purity water (3 mL) and methanol (7 mL) in a 100 mL round bottom flask with a stirring bar. The mixture was stirred, while the monomer AAm (2 g, 28.15 mmol) was added and the solution was bubbled with argon. After 15 min, CuBr (64 mg, 0.45 mmol) was added and the solution was bubbled for further 5 min, until all CuBr is dissolved. Before the reaction could be started, the functionalized gold substrates were placed in the reaction flask and flushed with argon for 5 min. Afterwards, the reaction solution was transferred by a syringe to the funnel and purged with argon for further 5 min. With the addition of the solution to the reaction flask the polymerization was initiated and was continued for a definite reaction time in a closed system under argon at room temperature. To stop the polymerization, the samples were pulled out of the solution and they were rinsed and washed with high purity water and ethanol and dried with a clean and dry nitrogen gas flow.[6]

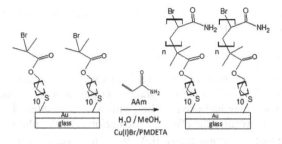

Figure 4.6: Scheme of the atom transfer radical polymerization (ATRP) with AAm

For the observation of the time dependency (**polymerization kinetics**) of the polymer brush growth, polymerizations for different times were carried out at ambient temperature (~23 °C) and constant solvent composition starting from pure initiator SAMs. Polymer brushes with different **graft densities** were prepared on mixed SAMs, containing an initiator and a dummy initiator moiety, for a reaction time of 60 min and at ambient temperature. At least, polymer brushes with different **cross-linking densities** starting from pure initiator SAMs were prepared for a reaction time of 60 min and at ambient temperature. And this time the feed of the cross-linker (bisAAm) in the reaction solution was varied.

4.3 Characterization

4.3.1 Contact Angle (CA) Measurements

Since the gold substrates were modified with different kinds of SAMs and afterwards polymerized, an interesting non-invasive method, the CA measurement, for the observation of the surface composition was performed. The contact angles of various surfaces were measured using a Dataphysics OCA-15 equipped with a digital camera under ambient conditions. For the sessile drop technique, in order to obtain the static contact angle, a 2 µL drop of Milli-Q water was dispensed automatically by a syringe (0.5 µL/s). Immediately after the water droplet is deposited on the surface, a picture is manually taken and was measured using a drop shape analysis software. For the dynamic contact angle, the advancing and receding angles, were measured by placing a 2 µL drop on the surface, while the needle was still in contact with the drop. A constant change of the volume (1 µL) was applied by sucking and releasing the volume for 3 times. Always the average of the second maxima and minima was analyzed. Both kinds of measurements were performed three times per sample at different locations, so the values are reported as the average and its standard deviation.

4.3.2 Ellipsometry

Ellipsometric measurements are routinely employed to measure the nanoscale thickness of films. The polyacrylamide-grafted samples were measured directly after the synthesis, when they had been washed with water and ethanol and then dried with a nitrogen gas stream. The ellipsometric thickness of the film was measured in air with an alpha-SE® ellipsometer (J.A. Woollam Co., Inc.) at three different incidence angles (of 65°, 70° and 75°) with a wavelength between 380-900 nm. The ellipsometer determines a sum parameter of all the material on the gold surface. In this case the thickness was determined via analysis of a two-layer model (1: background, 2: polymer). The constants of gold substrates were derived from ellipsometric measurements with a Cauchy layer model provided with the instrument software conducted on 3 different spots on a bare gold substrate. The recorded substrate constants of the "background" were used to identify the correct thickness of the attached initiator or grown polymer. The refractive index of poly(acrylamide) was estimated to be 1.45 at a wavelength of 632.8 nm.[12] Thus, the ellipsometric thickness for each sample was independently measured at three different locations and is reported as the average and its standard deviation, whereas the standard deviation represents the error for each value.

4.3.3 Fourier Transform Infrared (FTIR) Spectroscopy

Infrared spectra were used for the analysis of the composition of PAAm brushes. The FTIR instrument used was a Bruker IFS 66 with a mercury cadmium telluride quantum detector. This detector is faster than a thermal detector. For low noise detection, it has to be cooled with liquid nitrogen. The measurement was done in reflection mode. Background spectra were collected by scanning a cleaned gold wafer with a SAM of a deuterated thiol. The spectrometer software, OPUS, automatically subtracts the background from the sample. All measurements were recorded at an incident angle of 80° and under vacuum in order to avoid additional absorptions through the atmosphere (e.g. water or CO_2). 2000 scans were taken automatically and accumulated by the software. The final FTIR diagram was generated with Origin 7.5. The baseline was defined in a standard procedure manually. It has to be taken in account that the introduced error is larger when the peaks are small, because the manual definition of the baseline is not trivial especially for these small signals.

4.3.4   Atomic Force Microscope (AFM)

The morphologies of PAAm brushes in a dry and wet state were measured by a Multimode AFM, produced by VEECO. The advantage of an AFM is that the morphology of surfaces can be investigated in nanometer scale range as well as the physical properties of nanometer sized samples.

AFM images, for the following thickness and roughness analysis, were taken in Tapping Mode (TM) at room temperature (23°C) with a cantilever (PDNISP; stainless steel), which has a resonant frequency of 35-65 Hz and a spring constant of 100-300 N/m. The final analysis of the AFM images was done with a program called Nanoscope 531r1. In a first step the height images had to be flattened by the option "Plane Fit Auto" in the order 1 or 2 (2 was only used for images measured in water). Then, one area of the images with edges was defined and flattened again. As a second step, the analysis of roughness and thickness could be started. For the *roughness* analysis, the option "roughness analysis" was selected and the given results (RMS and RA values) of three defined areas were collected. The average RMS roughness value is the standard deviation of all pixel values from the mean pixel value. RA represents the standard deviation of a pixel value from the mean plane.

The *thickness* analysis could be performed in different dimensions 1D, 2D or even 3D. In the case of the thickness analysis in 1D, the option "section" was selected and three cross sections of the height image were sequential selected and the displayed thickness in nm ("vert distance") was collected. For the thickness analysis in 2D, the option "step height" was selected and the analysis part of the image had to be verified. First, an area of the height image was chosen by adjusting the white vertical lines. Afterwards, the cross section of the height image was adjusted by moving the two green and red arrows, whereas each colored pair replaces a flat area. Then the step height was displayed and collected. In the end the option "bearing" was selected for the thickness analysis in 3D. Two cursers were placed on the maxima of the bearing distributions, whereas the difference between these cursers represents the depth of the step.

AFM *force curves* were taken in Contact Mode (CM) with MLCT cantilevers from Bruker. The measurements were performed in air as well as in water, with spring constants in the range 0.10 N/m and tip radii of 20-60 nm following the publication of Spencer et al.[12] With the help of a newer version of Nanoscope v720, the measured force curves were transformed in single data sets for further analysis with Microsoft Excel 2010. The spring constant was evaluated with the thermal noise method (Asylum MFP-3D). The deflection sensitivity of the optical beam detection system was calibrated with a silicon wafer. The tip radius was evaluated using NioProbe calibration samples.

5   Results and Discussion

The analysis of PAAm brushes is divided into two main parts: first, the characterization of polymer brushes will be reported followed by the analysis with the AFM.

5.1   Polymer Brushes

The resulting conformation and constitution of polymer brushes can be influenced by a variety of factors before and during polymerization via SI-ATRP. In the beginning of this study the time dependency of the growth of polymer brushes of AAm was analyzed to select a considerable thickness for the following analysis of the influence of cross-linking and grafting density. PAAm brushes on gold surfaces were prepared, following the formation of initiator SAMs of MUBiB on gold surfaces. The characterization of PAAm brushes was performed with ellipsometry, FTIR spectroscopy and CA measurements.

5.1.1   Polymerization Kinetics

Figure 5.1: Schematic representation of the polymer brush growth

The polymerizations for different times were carried out at ambient temperature (~23 °C) and constant solution composition starting from pure initiator SAMs, as shown in Figure 5.1. The formation of a uniform MUBiB (initiator) monolayer on gold wafers was investigated by measuring the contact angle of the modified substrate. For the MUBiB-modified substrates an average contact angle of 81 ± 1° was obtained, which can be related to a considerable hydrophobic effect of MUBiB. The obtained water contact angle is in good agreement with previously reported values (82°).[6] The presence of an organic thin layer was additionally confirmed by ellipsometry, whereas an average ellipsometric thickness of 1.8 ± 0.3 nm was measured. These thickness values are consistent with previously observed values and the values predicted by bond length calculations (1.5-2.0 nm).[23] Afterwards, the initiator SAMs were used for polymerizations of different times, starting from 5 min up to 240 min.

The presence of a PAAm layer on the surface can be inferred from a definite change of the water contact angle (Figure 5.2). In total, the contact angle of

PAAm brushes was less than 24°, which is significantly lower than the contact angle of the initiator SAM (81 ± 1°).

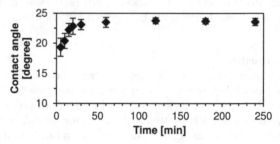

Figure 5.2: Dependence of the water contact angles on the polymerization time of PAAm

In Figure 5.2 the variation of the contact angle of water as a function of the polymerization time of PAAm is plotted. The data shows a change of the contact angle of PAAm brushes, which increases gradually from 18° to 24° with polymerization time. After a polymerization time of 20min the contact angle stagnates around 24°. The hydrophilic nature of the grafted surface is due to strong hydrogen bonds formed between amide groups of PAAm and water.

The film growth rate of PAAm brushes was investigated by ellipsometry. In Figure 5.3 the thickness is plotted as a function of the polymerization time. PAAm brushes were rinsed with water and ethanol and blown dry in nitrogen gas flow, to ensure similar wetting conditions.

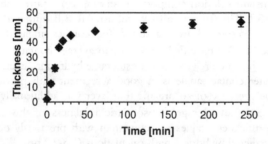

Figure 5.3: Thickness of PAAm as a function of polymerization time characterized by ellipsometry [Reprinted from Covalently cross-linked poly(acrylamide) brushes on gold with tunable mechanical properties via surface-initiated atom transfer radical polymerization, **2013**, *49*, Lilge, I., Schönherr, H., 1943-1951., Copyright 2016, with permission from Elsevier.]

Figure 5.3 shows a linear increase in polymer brush thickness for short reaction periods. After a polymerization time of 60 min the polymer brush growth stagnates. The non-linearity in the time dependence indicates a termination due to termination from radical combination, the dropping of monomer concentration or the loss of active catalyst. To proof the synthesis strategy, the measured thicknesses by ellipsometry were compared with the results reported by Liu et al, as shown in Figure 5.4.[6]

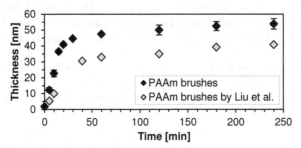

Figure 5.4: Comparison of the polymerization kinetic determined by Liu et al. with my results

As the Figure 5.4 shows, the thickness trend measured with the ellipsometer of the brushes is consistent with the thickness trend of the polymer brushes of the same monomer reported by Liu et al., whereas the SI-ATRP of AAm slowed at long reaction times. Also, other studies observed a rapid growth at early stages of the polymerization and a termination after about 60 min.[36] But in total, the values of my prepared brushes are generally greater than the ones determined by Liu et al.

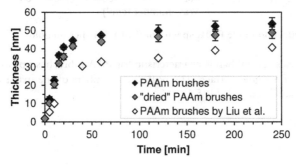

Figure 5.5: Comparison of the polymer brushes in a wet and dry state with the results determined by Liu et al.

An additional drying procedure overnight in an oven at 37 °C decreased the brush thickness only about 3-5 nm ("dry results" in Figure 5.5). A Soxhlet extraction (in Milli-Q water, at 110 °C, in argon atmosphere and in a dark fume hood), which is used for the removal of any physisorbed monomeric and poly(acrylamide), also revealed a small decrease of the brush thickness (around 3 nm) of the PAAm bushes polymerized for 60 min. The reduction in thickness may be attributable to the fact that physisorbed AAm monomer or chains, which are linked to oxidized sulfur, were desorbed during the extraction process. The overall higher values may be also due to the good conditions during the polymerization, as purer Cu(I)Br and an absence of Cu(II)Br or even a better closed reaction system, where the oxygen is excluded.

FTIR spectra of PAAm brushes were recorded and analyzed to verify the chemical composition of polymer brushes upon different reaction times. First a figure of thick PAAm brushes is shown to link the major absorptions of the spectra to vibrations of the polymer.

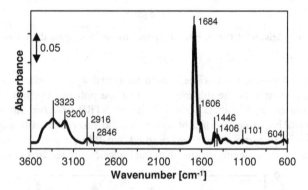

Figure 5.6: Reflection mode FTIR spectra of PAAm brushes on gold

Figure 5.6 reveals several bands characteristic for PAAm brushes polymerized from initiator SAMs on a gold surface. The peak wavenumbers of the major absorptions in the spectra are listed in Table 5.1.

Table 5.1: Peak assignments and wavenumbers for IR-spectra of PAAm on gold

vibration	origin	PAAm [cm^{-1}]	
ν_{as}(NH)		3323	*Fermi-resonance*
ν_s(NH)	amide II band	3200	*Fermi-resonance*
$\nu_{as}(CH_2)$	*polymer chain*	*2916*	
$\nu_s(CH_2)$	*polymer chain*	*2846*	
ν(C=O)	amide I band	1684	
δ(NH$_2$)	amide II band	1606	
ν(C-N) + ν(C-C) + δ(N-H)	amide II band	1502	cross-linker
ν(CN) + δ_s(CH$_3$)		1446 - 1406	
ν(C-H) + δ(C-N-H)	amide II band	1310 - 1250	*very low*
δ(NH$_2$) *rock*		1150	
δ(NH$_2$) *wag*		~700	

Characteristic absorption bands of PAAm brushes are around 3323, 3200, 1584, 1611 cm^{-1}. The broad absorption bands at 3323 and 3200 cm^{-1} are attributed to the stretching vibrations of the amine functionality. These are followed by smaller broad CH$_2$-streching vibrations, which are mainly due to vibrations of the polymer chain. The more distinct signals at 1684 and 1606 cm^{-1} are assigned to the amide I and amide II vibrations, respectively. Similar FTIR spectra were reported by van der Mei et al.[83]

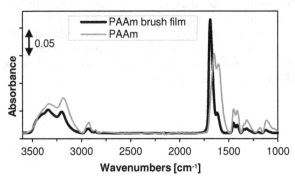

Figure 5.7: ATR-FTIR spectra of PAAm purchased from Aldrich (dotted) compared to reflection mode FTIR spectra of PAAm brushes

Attenuated total reflection (ATR)-FTIR spectra confirmed the chemistry of polymer brushes grown from a gold substrate as shown in Figure 5.7. For grafted PAAm brushes a very large decrease in the intensity of all peaks, especially in the region of shorter wavelength, was observed, which is attributable to the two different FTIR methods. The relationship between the carbonyl peak and the $\delta(NH_2)$ vibrations additionally changed and according to the previous mentioned fact of another method, the orientation of the bonds affects the peak intensities recorded via reflection mode FTIR.

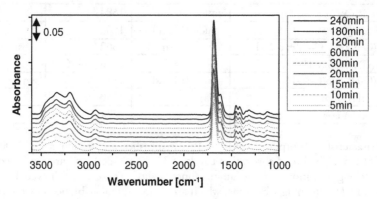

Figure 5.8: Reflection mode FTIR spectra of PAAm brushes with different thicknesses due to the variation of the polymerization time

The FTIR spectra displayed in Figure 5.8 for PAAm brushes polymerized for different times are very similar. A definite increase in intensity of all absorption bands with brush thickness is easily observed. A correlation of the intensity of the carbonyl peaks with the thickness should result in a linear dependency. For an easier analysis of this relationship, the area of the intensive peak of the carbonyl vibrations and the NH_2 vibrations was calculated.

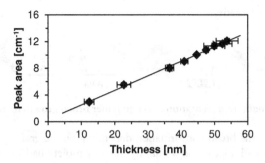

Figure 5.9: Linear relationship between the peak areas of the C=O vibrations and the polymer layer thickness

In Figure 5.9 the peak area of the carbonyl peak is plotted against the thickness of polymer brushes measures by ellipsometry. It is easily seen, that the area of the carbonyl peak is in a linear dependency to the polymer layer thickness, with an intersection through zero. The enclosed linear fit verifies this relationship.

5.1.2 Grafting Density

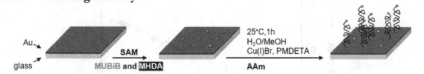

Figure 5.10: Schematic representation of the polymer brush growth on mixed SAMs

For the analysis of the brush conformation depending on the grafting density mixed SAMs containing an initiator and a dummy initiator moiety had to be prepared, as shown in Figure 5.10. Since the formation of mixed SAMs could have a non-ideal behavior, i.e. the surface composition of the mixed SAM could differ from the solution composition, the binary SAMs had to be analyzed beforehand. In this case the contact angles were measured to study the wettability of the binary SAMs and the surface composition was calculated afterwards by using the Cassie equation and the Israelachvili equation.

The pure SAMs, containing only the initiator (MUBiB) or 16-mercaptohexadecanoic (MHDA), showed a static contact angle for water of 81 ± 1° and 10 ± 1°. SAMs obtained from pure MHDA or MUBiB solutions had respectively a very low (hydrophilic) and intermediate contact angles. This means for the SAMs with pure MHDA, that the water spreads nearly completely on the surface and a contact angle of 10 ± 1° is obtained as shown in Figure 6.11a. This result is consistent with the reported values by Weiss et al.[77]

a)   b)

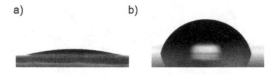

Figure 5.11: Contact angle of a) a pure MHDA SAM (10 ± 1°) and b) a pure MUBiB SAM (81 ± 1°)

Later-on, the average of all values of the static and dynamic contact angle measurements was calculated, together with the standard deviation. Because SAMs obtained by adsorption from mixed thiol solutions, lead to a surface composition where one of the components is preferentially absorbed on the surface, which is depending on the chain length, chemistry and surface affinity of the thiol. All measured values were plotted into a diagram (Figure 5.12) with the fraction of MUBiB in the solution on the x-axis against the contact angle [degree] on the y-axis, to show qualitatively the amount of initiator in the mixed SAM. The standard deviation represents the error for each value.

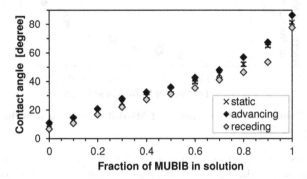

Figure 5.12: Comparison of the contact angles with the fraction of MUBiB in the solution

Figure 5.12 shows a linear change of the contact angle until a percentage of MUBiB in the solution is ca. 80%. Afterwards, the coverage of the MUBiB on the surface is high enough to have an effect on the contact angle, so that the linear change increases until the maximum contact angle is reached for pure MUBiB (Figure 6.11a). For all values the static contact angle lays in between the dynamic contact angles. Mixed SAMs show contact angles in between the values measured on pure MHDA and pure MUBiB SAMs. It is worth to notice, that the error for the low static contact angles is larger than for the higher ones. Measuring very low contact angles is indeed more difficult, since the drop shape has to be estimated, while the drop spreads on the surface. The error for the concentration in the solution, calculated for a thiol solution of 5 mL, was negligible as also the error of the dynamic contact angles, which was lower than 0.5°.

After the contact angle of each SAM was observed, it was necessary to calculate the real surface composition for further investigations. In this case the fractional surface coverage was calculated from the static contact angles with the Cassie equation 4.2 and the Israelachvili equation 4.3. Figure 5.13 represents these values with the fraction of MUBiB on the surface against the fraction of MUBiB in the solution.

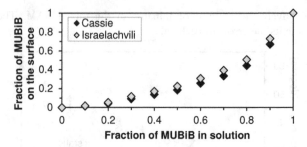

Fraction of MUBiB in solution

Figure 5.13: Comparison of the fraction of MUBiB in solution with the fraction of MUBiB on the surface

Both equations were used to estimate the MUBiB coverage on each sample giving similar results, despite the differences in the assumptions of the two theories. Figure 5.13 shows that the MHDA fractional coverage of the surface is always higher than expected for the mixed SAMs. The error of the contact angle determination (the standard deviation) was afterwards used for the error propagation performed with the Cassie and Israelachvili equations, and again the determined error was negligible. For the following polymerizations, only mixed SAMs with appreciable initiator coverage were used, i.e. the SAMs obtained after depositing the ATRP initiator MUBiB from the solution with the concentration of 30, 40, 50, 60, 70, 80, 90, and of course 100 percentage. Afterwards the change of the chemical composition of the mixed SAMs was analyzed by examining the FTIR spectra, as shown in Figure 5.14.

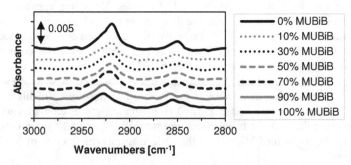

Wavenumbers [cm⁻¹]

Figure 5.14: Reflection mode FTIR spectra of mixed SAMs prepared with different fractions of initiator (MUBiB) in solution [Reprinted with permission from Lilge, I., Schönherr, H. Control of Cell Attachment and Spreading on Poly(acrylamide) brushes with varied grafting density. *Langmuir* **2016**, *32*, 838-847. Copyright 2016 American Chemical Society.]

The spectra in Figure 5.14 shows all bands expected based on the structure of the adsorbed thiols. Both thiols show the same two absorption peaks at ~2920 cm^{-1} and ~2859 cm^{-1}. These wavenumbers represent the asymmetric and symmetric stretching vibrations of the methylene groups. The intensity of these two peaks increases with the fraction of MHDA in the mixed SAM. With increasing amount of MHDA during formation of these mixed SAMs also a shift to lower frequencies as well as decreased peak widths are observed. The shift to lower frequencies indicates weaker hydrogen bonds due to the geometrical constrain and the presence of neighbor atoms in the SAMs with MHDA. A shoulder at ~2910 and ~2845 cm^{-1} is observed in the spectra of a pure MUBiB SAM. The absorption peak of the carbonyl vibrations of MUBiB was observed in the unlimited spectra at 1732 cm^{-1}. The observation of the chemical composition of the mixed SAMS was consistent with the results of a previous project. [84]

The polymerizations starting from SAMs with a considerable change of the initiator surface coverage were carried out for 60 min and at ambient temperature (~23 °C). Afterwards polymer brushes of PAAm on the mixed SAMs of MUBiB and MHDA were first characterized by their ellipsometric thickness. Figure 5.15

Figure 5.15shows the fraction of MUBiB in the solution on the x-axis against the average of all values of the measured ellipsometric thickness on the y-axis.

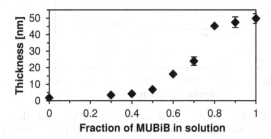

Figure 5.15: Thickness of PAAm as a function of the initiator (MUBiB) content in solution during SAM formation; characterized by ellipsometry [Reprinted with permission from Lilge, I., Schönherr, H. Control of Cell Attachment and Spreading on Poly(acrylamide) brushes with varied grafting density. *Langmuir* **2016**, *32*, 838-847. Copyright 2016 American Chemical Society.]

Considerably different thicknesses of polymer brushes were obtained starting with different MUBiB coverage. Because the polymers grafted on the substrate all have roughly the same degree of polymerization, the variation of the polymer brush thickness can be attributed to the difference in the density of the MUBiB grafting points on the substrate. A pure MHDA sample was polymerized as a reference, to ensure the usage as a dummy initiator. Ellipsometry measurements confirmed these assumptions, as the thickness of the thiol monolayer on the gold substrate (1.8 nm) was the same before and after polymerization. For very low initiator concentration (of MUBiB) in the solution, up to 50%, the ellipsometric thickness stays roughly constant, lower than 10 nm. The thickness of the brushes for 70% MUBiB in the solution shows an intermediate thickness. Approximately the same ellipsometric thickness of polymer brushes is reached for high concentrations of MUBiB in the solution, in the order of 50 nm. This is also the maximum of the thickness which can be reached after a polymerization of AAm for 60 min and the full initiator concentration on the surface.

Also for polymer brushes with different grafting densities FTIR spectra were recorded and analyzed to verify the chemical composition of polymer brushes upon different grafting densities.

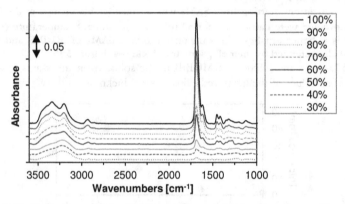

Figure 5.16: Reflection mode FTIR spectra of PAAm brushes with different thicknesses due to the content of initiator (MUBiB) in solution during SAM formation [Reprinted with permission from Lilge, I., Schönherr, H. Control of Cell Attachment and Spreading on Poly(acrylamide) brushes with varied grafting density. *Langmuir* **2016**, *32*, 838-847. Copyright 2016 American Chemical Society.]

Figure 5.16 shows a distinct increase of the peak intensity with the thickness of the polymer brush layer of PAAm obtained from mixed SAMs containing different fractions of MUBiB on the surface. Also in this case, the relationship between the polymer brush thickness and the intensity of the C=O peak with its shoulder of NH_2- vibrations is linear, as it can be seen in Figure 5.17. But it has to be taken into account, that the linear fit has no intersection through zero. This suggests that the polymer chains may be completely disordered and have a non-isotropic order due to the decreased grafting density. During the peak area analysis, a small shift to shorter wavelength, from 1684 to 1687 cm^{-1}, of the carbonyl peak was observed.

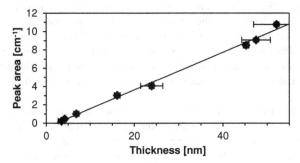

Figure 5.17: Linear relationship between the peak areas of the C=O vibrations and the polymer layer thickness [Reprinted with permission from Lilge, I., Schönherr, H. Control of Cell Attachment and Spreading on Poly(acrylamide) brushes with varied grafting density. *Langmuir* **2016**, *32*, 838-847. Copyright 2016 American Chemical Society.]

It has been shown, that the constitution of the polymer brush layer is linked to the thickness of the film. Another important fact is the connection with the fraction of the initiator on the surface. In this case the fraction of the initiator on the surface, estimated by the Cassie and Israelachvili equation is plotted against the ellipsometric thickness [nm] (Figure 5.18).

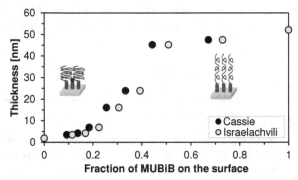

Figure 5.18: Comparison of the thicknesses [nm] of PAAm on mixed SAMs with MHDA with the fraction of MUBiB on the surface

Figure 5.18 shows a good agreement of the dependency of the ellipsometric thickness with the fraction of initiator on the surface obtained by the equations of Cassie and Israelachvili. In general, it is expected that the thickness increases with increasing initiator density on the surface. Instead, the polymer thickness remains approximately constant from very low fraction of the initiator on the surface up to 20%. Beyond 20% and up to 45% the thickness increases with initiator density until it reaches a plateau and the polymer thickness remains again approximately constant up to 100% initiator on the surface. This observation is in agreement with other reports.[21]

If other variables (temperature; time and monomer concentration) are constant, the change of the ellipsometric thickness for increasing values of initiator on the surface is largely due to the increase in brush density. In this case it is also assumed, that all substrates have the same polymerization rate, independent of the density of the initiator on the substrate. Therefore the increase of the polymer thickness can be attributed to the increase of the polymer grafting density on the substrate. At low grafting densities, the thickness is nearly independent of the initiator density. For these polymer brushes, the grown chains have little influence on each other because of the large distances between them. Hence, the surface anchored PAAm chains are in the mushroom regime, as indicated in Figure 5.18. At higher polymer grafting densities, the thickness increases with increasing grafting density, which is a signature of the brush behavior. In this case the initiator density has an influence on the polymer morphology, causing the polymer chains to undergo a mushroom-to-brush conformational transition above a critical surface density. This transition forces the active ends of the polymer chains to stretch up and away from the surface. The grafting density can be calculated from the following equation:[35,62]

$$\sigma = \frac{h \rho N_A}{M_n} \tag{5.1}$$

(ρ: density of PAAm (=1.302 g/cm³); N_A: Avogadro's number; M_n: polymer molecular weight)

While h can be accessed directly on the gradient sample, values of M_n on the substrate are not available. The final molecular weight of the polymer can either be yielded by a polymerization with "free-initiator" from solution or separation of the polymer brushes from the substrate. Afterwards, it would be determined by gel permeation chromatography (GPC). But in both cases for sufficient polymer weight analysis by GPC, the amount of the reactants is too high or the size of the gold substrate would be large, which makes the fulfillment of the polymerization conditions too hard and circumstantial.

With a certain initiator density on the surface (around 45%) the thickness of polymer brushes no further increases. This suggests that for higher initiator concentration only some out of the whole present initiators is used to start a polymer chain. Because the polymer chain backbone (~1.8-2.0 nm²) is much larger than the surface area of an initiator (~0.2-0.5 nm²), the growing polymer chains are sterically hindered, which leads to a blocking of some on the substrate present initiator molecules.[62]

The analysis of the surface composition, by contact angle measurements, of polymer brushes grown on gold surfaces with different initiator densities on the surface revealed a definite change with the grafting densities. In Figure 5.19 the fraction of the initiator on the surface is plotted against the contact angle [degree].

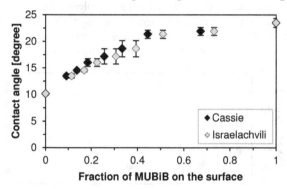

Figure 5.19: Influence of initiator density on the water contact angle

In Figure 5.19 an increasing contact angle with the grafting density of polymer brushes on the surface is observed. Starting with 10° of the reference sample of pure MHDA on the surface, which did not start the polymerization, the contact angle increases linearly with the fraction of the initiator on the surface. When a certain fraction of initiator is reached, around 45%, the polymer chains are in the brush regime and the contact angle is around 23°. The dependency of the contact angles on the grafting density can be due to an influence of the dummy initiator head groups, which still stick out of the thin polymer brushes of low grafting densities and as a result the contact angle decreases.

5.1.3 Cross-linking Density

Figure 5.20: Schematic representation of the polymerization with a cross-linking moiety

For the analysis of the polymer brush conformation depending on the cross-linking density, the polymerizations starting from pure initiator SAMs were carried out for 60 min and at ambient temperature (~23 °C). This time the fraction of a cross-linker, N,N'-methylenebis(acrylamide) (bisAAm) was varied in the polymerization mixture, as depicted in Figure 5.20. Afterwards polymer brushes of PAAm with different rates of bisAAm were first characterized by their ellipsometric thickness. Figure 5.21 shows the fraction of bisAAm in the solution on the x-axis against the average of all values of the measured ellipsometric thickness on the y-axis.

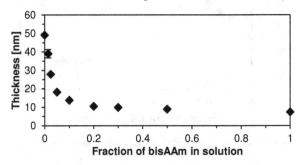

Figure 5.21: Thickness of PAAm as a function of cross-linking content (bisAAm) in solution characterized by ellipsometry [Reprinted from Covalently cross-linked poly(acrylamide) brushes on gold with tunable mechanical properties via surface-initiated atom transfer radical polymerization, **2013**, *49*, Lilge, I., Schönherr, H., 1943-1951., Copyright 2016, with permission from Elsevier.]

Considerably different thicknesses of the brushes were obtained by varying the concentration of the cross-linker (bisAAm) in solution. For high concentrations of bisAAm in the reaction solution, higher than 20%, the ellipsometric thickness stays roughly constant, around 10 nm. The polymer brush thickness starts to increase with lower fractions than 20% of bisAAm in the solution until a maximum thickness, which can be reached for a polymerization of pure AAm. The addition of the cross-linker to the polymerization solution leads to an increased tendency of early termination. The termination is caused by the cross-linking nature of these polymers, whereas the kinetics exhibit diffusion limitations. Consequential the termination kinetics strongly depend on the mobility of the reaction environment.[85]

Also for polymer brushes with different cross-linking densities FTIR spectra were recorded and analyzed to verify the chemical composition of polymer brushes with different fractions of bisAAm in the feed.

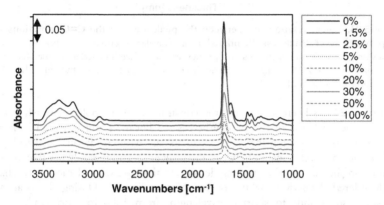

Figure 5.22: **Reflection mode FTIR spectra of PAAm brushes with different thicknesses due to the fraction of cross-linker (bisAAM) in solution** [Reprinted from Covalently cross-linked poly(acrylamide) brushes on gold with tunable mechanical properties via surface-initiated atom transfer radical polymerization, **2013**, *49*, Lilge, I., Schönherr, H., 1943-1951., Copyright 2016, with permission from Elsevier.]

The FTIR spectra displayed in Figure 5.22 for PAAm brushes polymerized with different fractions of cross-linker (bisAAm) in the polymerization solution. A definite decrease in intensity of all absorption bands with increasing cross-linker is observed, until only low intensive adsorption peaks can be observed. Besides the characteristic absorption peaks of PAAm, the additional appearance of a small absorption peak at 1502 cm[-1] can be attributed to the vibrations of the N-H bonding (amide II) band of the secondary amid in the cross-linker in the feed. The intensity of this peak increases with the increasing concentration of bisAAm in the solution and confirmed the incorporation of cross-linker in PAAm brushes.

A correlation of the intensity of the carbonyl peaks with the thickness should result again in a linear dependency and the according peaks were analyzed as discussed before. The table of all values is attached to this thesis.

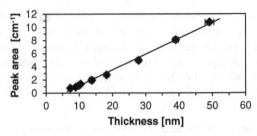

Figure 5.23: Linear relationship between the peak areas of the C=O vibrations and the polymer layer thickness [Reprinted from Covalently cross-linked poly(acrylamide) brushes on gold with tunable mechanical properties via surface-initiated atom transfer radical polymerization, **2013**, *49*, Lilge, I., Schönherr, H., 1943-1951., Copyright 2016, with permission from Elsevier.]

In Figure 5.23 has been shown that the constitution of the polymer brush layer is linked to the thickness of the film. Whereas the linear fit, implemented in Figure 5.23 shows no intersection through zero. This fact was already observed in Figure 5.17 for the linear relationship between the peak areas and the polymer thickness of polymer brushes with varied grafting density. In this case the polymer brush chains are disordered depending on the degree of cross-linking. During the peak area analysis, a small shift to shorter wavelength, from 1684 to 1689 cm^{-1}, of the carbonyl peak was observed. Thus, due to similar chemical structures of AAm and bisAAm, the degree of cross-linking of PAAm brushes could be tuned without significantly changing the overall chemical composition. Contact angle measurements revealed changes in wettability, respectively as a function of degree of cross-linking, as shown in Figure 5.24.

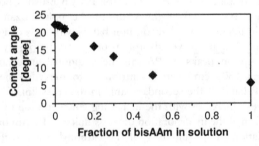

Figure 5.24: Influence of cross-linking density on the water contact angle of the polymer brush layer

Non-cross-linked PAAm brushes showed a contact angle of 22°, while addition of cross-linker led to a continuous decrease of the contact angle. The contact angle for bisAAm concentrations in the feed of >50% reached 5° and the surfaces is so called completely wetted. The definition of wetting is adopted from Bain et al., and means that water drops on a surface exhibit an irregular drop shape and a contact angle of less than 10°.

As already mentioned is the wettability of the resulting polymer brushes strongly affected by increasing cross-linking density, which is caused by a "bridging effect". The finite contact angles of grafted polymer chains on the surface are generated due to partial wetting and nonzero surface pressure. The lateral cross-linking of surface grafted chains is assumed to strongly hinder chain conformational freedom and thus inhibiting the bridging affect and causing an increase in wettability.[12,86] In Figure 5.24 is shown, that the bridging effect is completely absent in polymer brushes with low bisAAm contents (<1%) and then the brush surfaces showed greatly increased wettability compared to freely grafted brushes.

It is well known, due to Wenzel et al., that the wetting of solid surfaces is influenced by the roughness.[71] Consequently the significant decrease of the contact angle upon cross-linking will be compared with the following results of roughness measurements by AFM.

5.2 Polymer Brush Analysis by AFM

The variation of the degree of grafting density and cross-linking in surface grafted PAAm brushes was expected to strongly influence the surface and interfacial bulk properties of polymer brushes.[67,87,88] In order to investigate whether the brush morphology or lateral network of the grafted chains affects the characteristics of polymer brushes, a more detailed study by AFM was performed. These measurements revealed changes in morphology, wettability and mechanical properties respectively as a function of the grafting density and the degree of cross-linking.

Before the AFM analysis was started, it was necessary to grow polymer brushes in certain regimes; more precisely a polymerization was carried out from patterned initiator substrates, prepared by μ-CP as shown in Figure 5.25. Patterned polymer brushes are useful for the study of the thickness and thus responsive phenomena as the swelling behavior. The relationship between the conformation of the brushes on the surface in air or in water can be proven by measuring the thickness of PAAm brushes grafted on patterned SAMs.

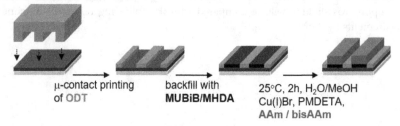

| μ-contact printing of ODT | backfill with **MUBiB/MHDA** | 25°C, 2h, H₂O/MeOH Cu(I)Br, PMDETA, AAm / bisAAm |

Figure 5.25: Schematic representation of the polymerization of patterned gold substrate

Therefore ODT μ-contact prints were performed on gold wafers, backfilled with the initiator and always polymerized with the previous measured samples by ellipsometer. ODT was chosen for the μ-contact prints due to its high surface affinity to gold. Because a previous project with the application in reversed order revealed a reduced amount of initiator molecules on the surface after an overnight treatment in an ethanolic ODT solution (1 mM). To ensure the persistence of the ODT molecules, a gold surface was imprinted by a stamp without any pattern and deposited in a MUBiB solution over night. The measured contact angles of the monolayers before (109.8 ± 0.8°) and after (107.2 ± 1.1°) revealed only a minimal change.

A fresh prepared ODT µ-contact print and the one, which was deposited in the MUBiB solution over night, were additionally polymerized. Because an important point to consider is the oxidation of thiols with the incorporation of initiator molecules in the monolayer as a result. Polymer molecules may also physisorb to the ODT monolayer. In this case the polymerized ODT monolayers (for 120 min) were analyzed by FTIR.

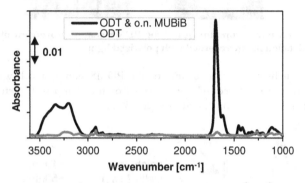

Figure 5.26: Reflection mode FTIR spectra of polymerized ODT monolayers (pure & with overnight treatment in a MUBiB solution

In Figure 5.26 the characteristic absorption bands of the PAAm are observed for the ODT sample with MUBiB treatment in comparison to the bare ODT monolayer. This means, that overnight in the initiator solution, some of the MUBiB thiols were incorporated into the monolayer and polymerized afterwards. But in the spectra of the bare ODT monolayer a very tiny carbonyl peak indicates the physisorbtion of monomer on the surface. When the peak area of the carbonyl peak (of ODT & o.n. MUBiB) was compared with the relation of the peak area and thickness observed with the results of the polymerization kinetics, the thickness of the persistent polymer brushes on the ODT monolayer was obtained. With this approach the polymer brush thickness lays around 12.3 nm. Additional ellipsometry measurements confirmed these results (11.8 ± 0.1 nm).

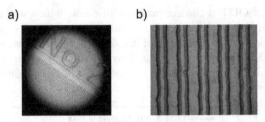

Figure 5.27: Light microscope images of the PDMS stamp a) overview of the stamp and b) more detailed image recorded with polarized light

In Figure 5.27 light microscope images of the PDMS stamp, which was used for polymerizations are displayed. The dimensions of the stamp were determined by SEM measurements as demonstrated in Figure 5.28.

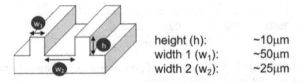

height (h):   ~10μm
width 1 (w_1):   ~50μm
width 2 (w_2):   ~25μm

Figure 5.28: Schematic of the PDMS stamp with estimations of the dimensions out of SEM measurements

With the determined dimensions and the verification of a relative stable μ-contact print formation without loss of ODT during the deposition in the initiator solution, the polymerizations were performed as reported before.

5.2.1 Surface Morphology

The contact angle measurement of the polymer brush surfaces revealed an increasing wettability with polymerization time, increasing grafting density and decreasing cross-linking density. In this case, a more detailed study of the surface morphology of PAAm brushes is performed by AFM roughness measurements. For the surface morphology analysis of polymer brushes, the height images taken by TM AFM in the dry state were used. Each time the values of RMS and RA roughness for three squares out of a 5 x 5 μm scan area were collected, as shown in Figure 5.29.

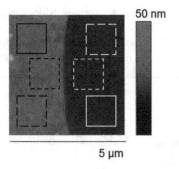

Figure 5.29: TM-AFM height image of PAAm brushes (240 min; 100% MUBiB; 100% AAm), to demonstrate the roughness analysis

The results of the roughness measurements of the polymer brush surface morphology exhibited definite differences between the ODT layer and the polymer brush layer. In contrast, the roughness of the ODT SAM is nearly constant for all samples (RMS= 3.3 ± 0.1 nm; RA= 2.7 ± 0.1 nm). This is consistent for PAAm brushes with varied polymerization time, grafting density and cross-linking density.

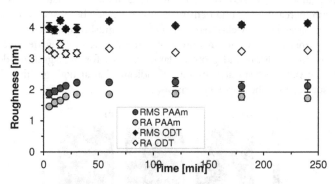

Figure 5.30: Roughness of PAAm brushes compared to the ODT monolayer as a function of the polymerization time; characterized by AFM

In Figure 5.30 the roughness of PAAm brushes is plotted as a function of the polymerization time. The roughness of the polymer brushes increases minimal with time in contrast to the roughness of the ODT layer. For thick polymer brushes of polymerization times longer than 60 min the roughness is nearly constant. In comparison to the ODT monolayer, the PAAm brush surfaces showed a uniformly smooth morphology and presented an average RMS roughness of 2.2 ± 0.1 nm. Spencer et al. reported an average RMS roughness of 0.4 nm.[12] The difference is due to the underlying gold substrate, which exhibits already a greater surface roughness than silicon, which was used by Spencer and his coworkers.

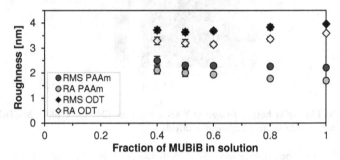

Figure 5.31: Roughness of PAAm brushes compared to the ODT monolayer as a function of the initiator (MUBiB) content in solution during SAM formation; characterized by AFM [values in Table 9.10]

AFM height images are shown in Figure 5.32 indicating that the roughness decreases with increasing grafting density, as it can already be observed by eye. The small clusters are likely well-separated polymer mushrooms formed on the extremely low initiator coverage region. For intermediate grafting densities, these clusters begin to disappear and grow to large patches for high grafting densities. The pretty low values of surface roughness indicate that at high grafting densities the PAAm surface morphology is very smooth.

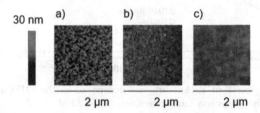

Figure 5.32: TM-AFM height images of representative surface morphologies recorded for PAAm brushes polymerized on sputtered gold with SAMs composed of a) 40%, b) 60% and c) 100% initiator in solution during self-assembly

Thus, the lateral variation in chain density has an influence on the surface morphology, which is consistent with the observations by Bohn et al.[89] In connection with the decreased surface roughness of PAAm brushes with increasing grafting density, the contact angle increases, as reported before. The average roughness values of polymer brushes are for that reason plotted together with contact angle against the fraction of initiator (MUBiB) in solution (Figure 5.33).

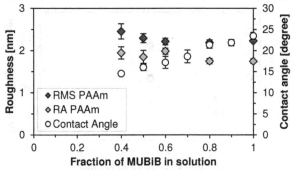

Figure 5.33: Roughness of PAAm brushes compared with the contact angle as a function of the initiator (MUBiB) content in solution during SAM formation

The contrary trend of the surface roughness and the contact angles is explained with Wenzel's equation, which states that surface roughness leads to an increase of the contact angle for smooth contact angles less than 90°.

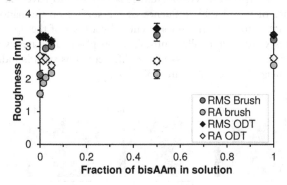

Figure 5.34: Roughness of PAAm brushes compared to the ODT monolayer as a function of the cross-linker (bisAAm) content in solution during polymerization; characterized by AFM

The addition of cross-linker led to a roughening of the dry polymer brush surface as displayed in Figure 5.34. The spatial variations in the cross-linking density during polymerization lead to a clustering of the growing chains. When the surface roughness measured by AFM is compared with the already mentioned contact angle dependency of polymer brushrs with different fractions of bisAAm in the feed, an opposite reliance is observed, as demonstrated in Figure 5.35.

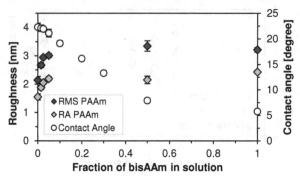

Figure 5.35: Roughness of PAAm brushes compared with the contact angle as a function of the cross-linker (bisAAm) content in solution during polymerization

This contrary trend of surface roughness and contact angle is again explained by Wenzel's equation. For the later on planned force curves, the roughness of the underlying substrate has to be considered. It is essential to have flat and smooth surfaces, because of the possible influence of the surface on the nanoindentations due to introduced distortions in the force curve by a twist of the tip.[74] In this case, polymer brushes were additionally synthesized on glass substrates with template-stripped gold. The surface roughness of the different substrates is indicated below with TM-AFM height images in Figure 6.37.

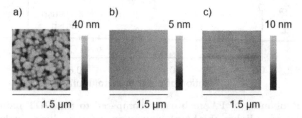

Figure 5.36: TM-AFM height images of representative surface morphologies recorded for a) sputtered gold, b) silicon and c) template-stripped gold

A great difference of the roughness values for sputtered gold and template-stripped gold substrates is already observed by eye in Figure 6.37. According to the TM-AFM height images, the corresponding micrographs (a section of the surface) are shown in Figure 5.37 to demonstrate the differences in roughness.

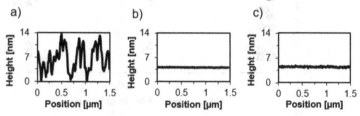

Figure 5.37: Micrographs to the corresponding TM-AFM height images of a) sputtered gold, b) silicon and c) template-stripped gold

The small difference of the silicon substrate and the template-stripped gold surface depends on the accurateness of the preparation as the cleanness of the silicon substrates, the sputtering rate of gold, and the stripping of the Si-Au-epoxy glue-glass "sandwich" in the end. The average roughness values, RMS and RA, are collected in Table 5.2.

Table 5.2: Comparison of roughness values (RMS & RA) of sputtered gold, silicon and template-stripped gold

substrate	RMS [nm]		RA [nm]	
	average	S	average	S
sputtered gold	5.0E+00	5.0E-01	4.0E+00	3.7E-01
silicon	4.8E-02	7.0E-03	3.9E-02	6.0E-03
template-stripped gold	1.6E-01	1.9E-02	1.3E-01	1.5E-02

Taking these results into account, the template-stripped gold surfaces provide an extremely smooth and clean gold surface for the realization of AFM studies on SAMs and polymer brushes. Therefore, it provides an excellent surface for the polymer study of the polymer brush surface morphology. In this case the average roughness values of PAAm brushes on sputtered gold are compared to the ones obtained from PAAm brushes grown on template-stripped gold.

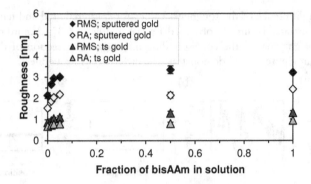

Figure 5.38: Roughness of PAAm brushes grown from sputtered and template-stripped gold as a function of the cross-linker (bisAAm) content in solution during polymerization

In Figure 5.38 similar trends of the surface roughness of PAAm brushes grown from sputtered and template-stripped gold as a function of the cross-linker in solution during polymerization are observed. The smooth surface of PAAm brushes has a RMS value of 0.9 nm, which is in the range of the reported values by Schouten et al.[62] Additionally TM-AFM height images of representative surface morphologies, which indicate the obtained roughness, are shown in Figure 6.40. The comparison of the height images revealed, that brushes with different amounts of cross-linker in the feed during polymerization have different surface morphologies.

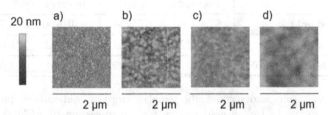

Figure 5.39: TM-AFM height images of representative surface morphologies recorded for a) 100% bisAAm, b) 5% bisAAm, c) 2.5% bisAAm and d) 0% bisAAm (100% AAm) brushes [Reprinted from Covalently cross-linked poly(acrylamide) brushes on gold with tunable mechanical properties via surface-initiated atom transfer radical polymerization, **2013**, *49*, Lilge, I., Schönherr, H., 1943-1951., Copyright 2016, with permission from Elsevier.]

5.2.2 Thickness

The patterned substrates were already polymerized together with the previous reported non-patterned substrates. Under these circumstances similar polymerization conditions are assumed. In this case, the thickness dependencies characterized by ellipsometry are proven by AFM measurements as section, step height and bearing analysis of height images collected in Tapping mode.

Figure 5.40a shows a 3D height image of PAAm brushes (240 min, 100% MUBiB and 100% AAm), in which a very clear step from the ODT monolayer to the PAAm brushes is observed. For thin polymer brushes the corresponding AFM height images, shown in Figure 5.40b, were a valuable aid for the determination of the actual edge of the pattern.

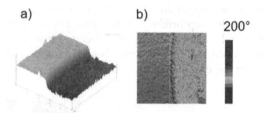

Figure 5.40: TM-AFM images of PAAm brushes (240 min; 100% MUBiB; 100% AAm) a) 3D height image and b) corresponding phase image in a different color scale

In Figure 5.41 - Figure 5.43, the three different methods for the determination of the thickness are figuratively explained. Starting with the section analysis, a line was drawn through the height image, perpendicular to the step, and three different cursers of the section were chosen for the analysis.

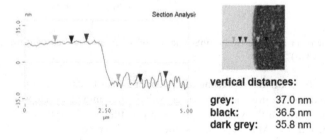

vertical distances:

grey: 37.0 nm
black: 36.5 nm
dark grey: 35.8 nm

Figure 5.41: Section analysis by NanoScope 5.3 of PAAm brushes (240 min; 100% MUBiB; 100% AAm)

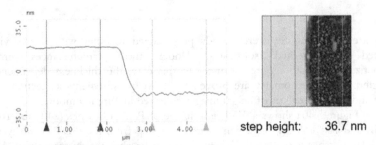

step height: 36.7 nm

Figure 5.42: Step height analysis by NanoScope 5.3 of PAAm brushes (240 min; 100% MUBiB; 100% AAm)

Figure 5.42 depicts the step height analysis; whereas the two green and red cursers each define a level and the step height was indicated. For the bearing analysis, in Figure 5.43, the green and red cursers respectively define a maximum of the bearing depth distributions. An additional green cursor is placed on the point of inflection of the bearing area. In the end the depth of the two levels in the height image is displayed.

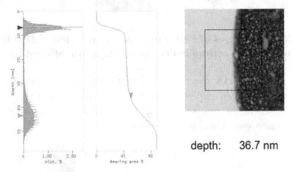

depth: 36.7 nm

Figure 5.43: Bearing analysis by NanoScope 5.3 of a PAAm brushes (240 min; 100% MUBiB; 100% AAm)

The following three figures represent the polymer brush thicknesses measured by AFM compared to the previously reported values determined by ellipsometry. A great difference between the values (decrease of one third) for the thickest polymer brushes estimated by AFM and ellipsometry is observed for all three cases, polymerization kinetics, grafting density and cross-linking density. However, the values of the thinner polymer brushes are in each case consistent for both types of analysis.

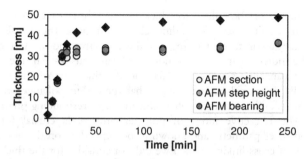

Figure 5.44: Thickness of PAAm as a function of polymerization time characterized by AFM compared to the results determined by ellipsometry

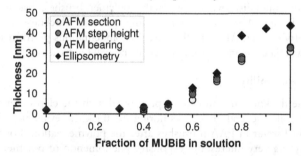

Figure 5.45: Thickness of PAAm as a function of the initiator (MUBiB) content in solution during SAM formation; characterized by AFM compared to the results determined by ellipsometry

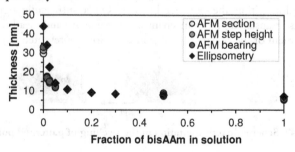

Figure 5.46: Thickness of PAAm as a function of cross-linking content (bisAAm) in solution characterized by AFM and compared to the results determined by ellipsometry

83

The difference of the values for thick polymer brushes measured by AFM and ellipsometry result either from additional PAAm brushes on the ODT monolayer as reported before or from a change of the refractive index. At the same time, it should be noted that also additional PAAm brushes on the stamped ODT monolayer can have a reducing effect on the thickness of the polymer brushes. Additionally, it should be mentioned, that these thin polymer brushes were very hard to detect and sometimes only visible in the corresponding phase images.

The refractive index of polymer brushes is crucial for the thickness estimation by ellipsometry and known to be dependent on the grafting density and the degree of cross-linking, which we did not consider for the thickness evaluation by ellipsometry.[90,91] The use of another refractive index (1.54) in our model, changed the estimated thickness only around 5 nm.[92,62] The influence of the surrounding humidity, which swells the polymer brushes, has to be taken into account. It also has an impact on the polymer thickness determination by ellipsometry or even AFM. In this case, nanoindentations performed by AFM shall indicate whether this fact has an influence and will be discussed later.

5.2.3 Wettability

So far, the thickness measurements performed with the ellipsometer and the AFM already revealed several drawbacks during the thickness evaluation due to humidity and residual water in PAAm brushes after the polymerization. For this reason, the wettability is a very interesting responsive phenomenon of polymer brushes, which experience swelling and collapse transitions in good and poor solvents, associated with large conformational changes of the polymer backbones. The apparent swelling/collapse caused by conformational changes, reflects an energy redistribution within the polymer brushes. On a sufficiently rigid substrate, the energy is translated into mechanical movement and leads to an internal rearrangement of the polymer brush conformation, represented in Figure 5.47.[45]

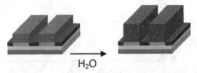

H_2O

Figure 5.47: Schematic representation of the swelling of patterned polymer brushes

Polymer brush layers, where the grafting distance is less than the dimension of the unperturbed chain, swell in good solvents. The dependency on the solvent quality can be adjusted by a temperature change. Suggesting that the grafted chains become highly hydrated when exposed to water, the overall change in thickness depends on the grafting density of polymer brushes. For low grafting densities, the polymer chains still have enough space to spread on the surface. With an increasing grafting density, the hydration layer around a polymer chain is well developed. Due to the minimization of repulsion interactions the polymer brushes extend to stretch away from the surface exhibiting a swelling ratio.[93] For higher temperatures, getting close to the LCST of the polymer, the hydrogen bonds are affected. Whereas the hydrogen layers undergo a partial degradation and the polymer chains are not well extended in water. In this case, the swelling ratio decreases.[26]

PAAm swells in water and consequently the thickness of the grafted polymer brushes increases strongly relative to dry conditions. The swelling ratio of PAAm brushes on patterned gold substrates was analyzed by TM-AFM measurements in water at room temperature (below LCST). Therefore, the substrates were exposed in an opened water system during the measurement and the swelling ratio is defined as:

$$Q = \frac{h_{swollen}}{h_{dry}} \qquad (5.2)$$

(Q: swelling ratio; h: film height)

As the swelling of surface-grafted polymer brushes is restricted to one dimension, the swelling ratio is a measure of the capability of brushes with varied grafting and cross-linking density to absorb water. The following two figures represent the dependency of the swelling ratio on the grafting and cross-linking density.

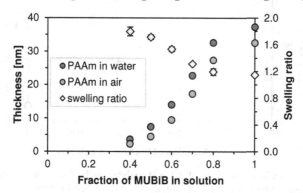

Figure 5.48: Swelling ratio of PAAm brushes with different grafting densities in water (dry and wet thickness determined by AFM)

PAAm brushes exhibit a high degree of swelling when immersed in water, whereas a correlation with the grafting density is observed in Figure 5.48. Already the ellipsometry measurements revealed a decrease of brush thickness with decreasing amount of present initiator on the surface, which is intensified for hydrated brushes. Despite this fact, the swelling ratio decreases with increasing amount of initiator on the surface, because the relative amount of swelling in comparison to the dry state is reduced for thick polymer brushes. This effect depends on the grafting density of polymer brushes, more precisely on the conformation of polymer brushes. For low grafting densities, the polymer chains have a lot of space to spread on the surface, whereas for high grafting densities the polymer chains are in the brush conformation and already extended and stretched away from the surface. In the brush conformation, the space in between the polymer chains is reduced, indicating that less water molecules can penetrate these denser brushes.

For polymer brushes with varied amounts of cross-linker during polymerization, the cross-linking of PAAm brushes lead to a decrease in the brush thickness with increasing bisAAm concentration. An absolute decrease in the dry polymer brushes is less pronounced than in the hydrated brushes, whereas the swelling ratio of PAAm brushes decreases with increasing bisAAm concentration. The swelling ratio of the PAAm brush thickness in the dry and hydrated state is plotted against the different feeds of bisAAm during polymerization in Figure 5.49.

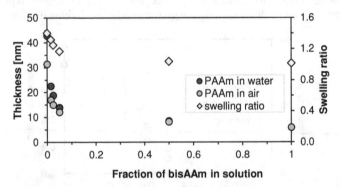

Figure 5.49: Swelling ratio of PAAm brushes with feeds of bisAAm during polymerization in water (dry and wet thickness determined by AFM)

This time the grafting density was kept constant for polymer brushes but the degree of cross-linking was varied, which has a great impact on the swelling ratio. Due to an increasing density of the network formation of PAAm brushes with increasing amount of cross-linker in the feed, the space in between the polymer chains is reduced and the swelling decreases. In contrast to these results, Spencer et al. observed much higher swelling ratios evaluated by means of ellipsometry, but with the same trend.[91] PAAm brushes exhibited in his case a swelling ratio of about 25.9 ± 1 (4.3 for 50% bisAAm; 2.2 for 100% bisAAm), which was comparable with AFM indentation measurements. The large response of PAAm brushes on water, high swelling ratio, suggests a number of practical applications, including creating surfaces with switchable adhesion, creating selective and switchable membrane and pores, and creating synthetic molecular motors.[45]

5.2.4 Mechanical Properties

The mechanical properties of polymer brushes are according to the previous mentioned responsive behavior on humidity an important aspect to analyze. In order to study how the mechanical properties of PAAm and covalently cross-linked brushes are influenced by the lateral cross-linking, nanoindentations by AFM were performed. The nanoindentations are fundamental measurements, whereas the polymer surface is non-destructively interrogated with a spherical surface probe to extract force data that can be correlated directly to physical properties of the polymer brush.[7]

But several issues concerning polymer brushes should receive attention. Besides the previously mentioned influence of surface roughness on the performance of force curves, the thickness of the analyzed polymer brushes has to be taken into account, because the brush thickness limits the scale on which the nanoindentation can be performed. If the indentations are too deep, the mechanical behavior would result from the coupled sample and substrate properties. A rule of thumb to avoid substrate effects would be to work at penetration depths less than one tenth of the total brush thickness.[74] In this case thicker brushes had to be synthesized on template-stripped gold by extending the polymerization time (120 min), before the nanoindentations on PAAm brushes were carried out. The ellipsometric thickness of PAAm brushes with varied cross-linking density synthesized on template-stripped gold for 120 min is in Figure 5.50 to the ellipsometric thickness of polymer brushes with varied cross-linking density synthesized on sputtered gold substrates.

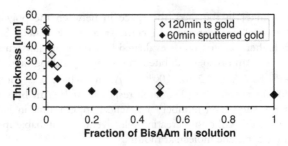

Figure 5.50: Thickness of PAAm brushes (polymerized for 60 and 120 min) as a function of cross-linking density characterized by ellipsometry

The measurements of the ellipsometric thickness of PAAm brushes with increasing degree of cross-linking revealed greater values for the longer polymerization period. Due to the introduction of the cross-linker the polymerization kinetics changed and for 60 min of polymerization the polymer brush growth does not stagnate. A change in thickness depending on the roughness of the underlying gold substrate was excluded by additional polymerizations of sputtered gold substrates with the template-stripped gold ones in the same reaction vessel, whereas both kinds of substrates revealed the same thicknesses measured by ellipsometry.

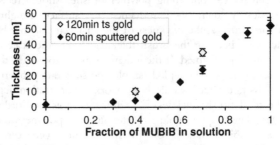

Figure 5.51: Thickness of PAAm brushes (polymerized for 60 and 120 min) as a function of initiator density characterized by ellipsometry

In Figure 5.51 the same relationship for polymer brushes with low grafting density is observed. With increasing polymerization time the brush thickness increased. But for PAAm brushes with pure initiator on the surface the thickness after 120 min is similar to the brush thickness of a polymerization of 60 min. These results are consistent with the previous mentioned dependency of the brush thickness on the polymerization kinetics.

The first nanoindentations of PAAm brushes with different degrees of cross-linking were performed with CM-AFM in air (ambient conditions). But all different polymer brushes were too stiff for the cantilever spring constant. The belonging curves of the indented PbisAAm and PAAm brushes are compared with the indentation of a bare silicon substrate in Figure 5.52.

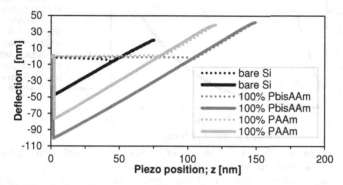

Figure 5.52: Nanoindentation curves on PAAm / PbisAAm brushes and Silicon in air

In Figure 5.52 the deflection varies linear in the same manner for all three different samples with the piezo position. In this case, these samples were exposed in an opened water system during the indentations by an AFM tip. The belonging curves of the indented PbisAAm still were as stiff as the bare silicon substrates for the utilized cantilever, as demonstrated in Figure 5.53. For this reason, the deflection sensitivity (41.37 nm/V) was determined on PbisAAm polymer brushes immersed in an open water cell.

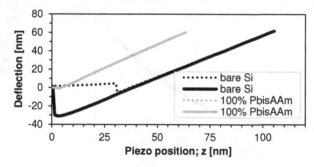

Figure 5.53: Nanoindentation curves on PbisAAm brushes and Silicon in water

Later on, polymer brushes with different grafting and cross-linking densities were indented in the same way and deflection-piezo position profiles were recorded for all samples, shown in Figure 5.54.

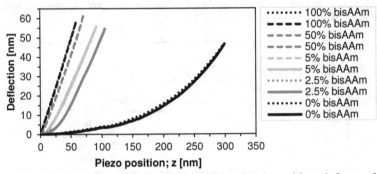

Figure 5.54: Raw nanoindentation curves of PAAm brushes with varied cross-linker (bisAAm) in the feed during polymerization [Reprinted from Covalently cross-linked poly(acrylamide) brushes on gold with tunable mechanical properties via surface-initiated atom transfer radical polymerization, **2013**, *49*, Lilge, I., Schönherr, H., 1943-1951., Copyright 2016, with permission from Elsevier.]

A change in slope of the different deflection-piezo position curves was observed accordingly to the concentration of cross-linker. The raw force curves of indentations with a smaller threshold are attached to this thesis as Figure 8.8. The nanoindentations of polymer brushes with different grafting densities also reviled a change in slope (Figure 5.55).

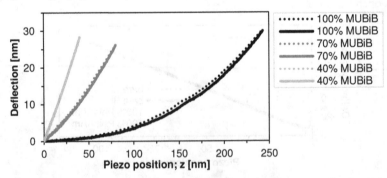

Figure 5.55: Raw nanoindentation curves of PAAm brushes with varied amount of initiator (MUBiB) in the mixed SAM

The influence of the polymer brush thickness should still be kept in mind during the evaluation of the raw force curves and their dependency on the cross-linking and grafting density. In both cases, with increasing amount of cross-linker or decreasing amount of initiator, the brush thickness is decreased and has an impact on the measured stiffness of polymer brushes.

The so-called raw force curves were transformed with Equation 4.7 and 4.8 into characteristic load-penetration depth curves. Therefore, the cantilever elastic constant was estimated with the thermal fluctuations method with the Asylum AFM, before and after the performance of nanoindentations on polymer brushes. First, the deflection sensitivity was determined three times on glass, secondly the average was calculated and filled in for the analysis of the thermal spectra of the cantilever. The values were very similar in both cases (Table 5.3), but one cantilever broke ("x") due to the several exchanging operations in and out of the different cantilever holders.

Table 5.3: Values of the different cantilever spring constants estimated on glass before and after nanoindentations on PAAm brushes compared to the nominal value supplied by the producer

cantilever	cantilever spring constant [pN/nm]	
	before	after
1.3E	138.31	134.94
1.9E	136.58	132.84
2.3E	151.40	x
2.6E	134.01	133.29
nominal value	100 (min.: 50; max.: 200)	

In comparison to the nominal values of the cantilever spring constant deviates about 30-50%, whereas typical deviations up to ± 200% are possible.[94]

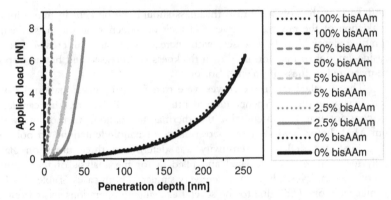

Figure 5.56: Applied load versus penetration depth curves for PAAm brushes with varied cross-linker (bisAAm) in the feed during polymerization [Reprinted from Covalently cross-linked poly(acrylamide) brushes on gold with tunable mechanical properties via surface-initiated atom transfer radical polymerization, **2013**, *49*, Lilge, I., Schönherr, H., 1943-1951., Copyright 2016, with permission from Elsevier.]

The load-penetration depth curves in Figure 5.56 correspond to typical force curves of an elastic solid and show qualitatively an increase in the elastic response with decreasing amount of cross-linker in the feed. The transformed curves of the smaller threshold are attached to this thesis (Figure 8.9). On the other hand, the elastic response of the curves of PAAm brushes decreases with decreasing grafting densities.

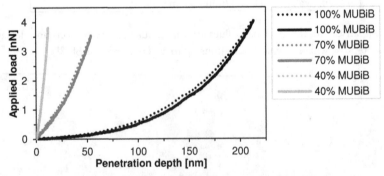

Figure 5.57: Applied load versus penetration depth curves for PAAm brushes with varied amount of initiator (MUBiB) in the mixed SAM

The force curves of polymer brushes with low grafting and cross-linking densities do not show a linear behavior after a penetration depth corresponding to the measured hydrated thickness. Thus, the substrate is not reached with these penetration depths and a new thickness of the PAAm brushes will be determined later on.

The Young's moduli of PAAm brushes and cross-linked brushes were evaluated according to Sneddon's model from the region of the force curves corresponding to penetration depths smaller 10 nm, in order to avoid substrate effects. The penetration depths smaller than 10 nm correspond to the dependency of the elasticity of polymeric films, which is demonstrated in Figure 5.58.

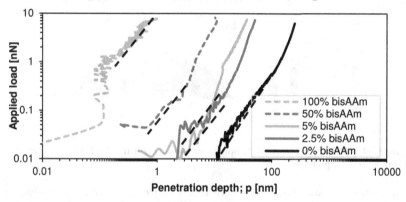

Figure 5.58: Logarithmic plot of applied load vs. penetration depth of PAAm brushes with varied cross-linker (bisAAm) in the feed during polymerization

With this knowledge, the Young's moduli were calculated using the slope of the applied load against penetration depth to the power of 3/2 curve, according to the transformation of the Sneddon's equation (Equation 4.9).

$$F = m * p^{3/2} \tag{5.3}$$

(F: applied load; m: slope; p: penetration depth)

$$m = \frac{4}{3(1 - \nu^2)} * E * \sqrt{R_{tip}} \tag{5.4}$$

(ν: Poisson ratio; R$_{tip}$: tip radius; E: Young's Modulus)

$$E = \frac{m * 3(1 - \nu^2)}{4 * \sqrt{R_{tip}}} \tag{5.5}$$

The resulting slopes of the curves were inserted in Equation 6.5 and the Young's Modulus was calculated. The used Poisson ratio of PAAm brushes, of 0.480 ± 0.012, was reproduced from Tracqui et al.[95] The tip radius was characterized with an algorithm based on the "Blind Tip Reconstruction" method with of SPIP of an 1 x 1 μm height image recorded from a Nioprobe with the Asylum AFM (Figure 5.59). The analysis revealed a tip radius of 15 ± 2 nm, which is much lower than the nominal value supplied by the producer (20-60 nm). But it has to be taken into account, that the small radius is estimated for the end of the tip (last 6 nm) and that the range of the tip radius, given by the producer, is very broad. The 3D view, displayed in Figure 5.59b gives a correct impression of the tip geometry.

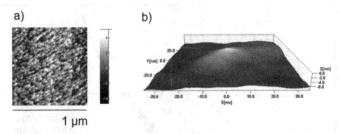

a)   b)

1 μm

Figure 5.59: CM height image of Nioprobe recorded with the Asylum AFM (a) and a 3D image of the tip shape.

In general, the elastic moduli (E) were found to lie in the range of 20-700 kPa according to the values reported by Spencer et al. and also to reported moduli for bulk PAAm hydrogels.[12] The calculated values of E are reported in Table 5.4.

Table 5.4: Calculated values of the elastic modulus (E) of several PAAm brushes with varied amount of cross-linker (bisAAm) in the feed or different grafting densities

bisAAm	m [nN/nm^{32}]	E [MPa]	
		high threshold	low threshold
1	8.60E+00	1.11E+03	8.87E+01
0.5	5.79E-02	7.47E+00	8.15E+00
0.05	2.49E-03	3.22E-01	8.89E-01
0.025	1.35E-03	1.75E-01	7.49E-01
0	5.74E-04	7.40E-02	7.06E-02

Initiator	m [nN/nm^{32}]	E [Mpa]
1	5.47E-04	7.06E-02
0.7	8.53E-03	1.10E+00
0.4	8.76E-02	1.13E+01

The Young's modulus of cross-linked brushes was more than 100-fold larger than native PAAm brushes. In the case of lateral cross-linking and reduced grafting densities, different Young's moduli have been obtained. While PAAm brushes behaved as a compliant film mainly constituted of water, a considerably stiffening of cross-linked PAAm brushes was observed upon increasing the concentration of bisAAm in the feed. In total, the Young's moduli could be precisely tuned for different brushes, by varying the amount of cross-linker in the formed brushes or even the grafting density. In Figure 5.60 the values of the elastic modulus are correlated with the corresponding swelling ratios of polymer brushes.

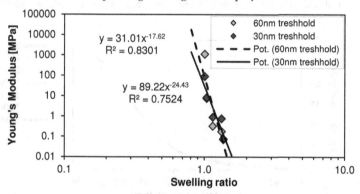

Figure 5.60: Comparison of the elastic (Young's) modulus with the observed swelling ratio of polymer brushes with varied amount of cross-linker in the feed (results of force curves of two different thresholds)

Comparing the ratio of the two curves in Figure 5.60 with the results obtained by Spencer et al. the scale with the exponent is a factor of 10 too high [Spencer: E $\propto$ Q$^{-1,57}$ or E $\propto$ Q$^{-2,25}$].[91] This is due to the difference of the determined swelling ratios of the polymer brushes. Because it should be taken into account that the tip-brush contact point in the applied load vs. penetration depth curves of less cross-linked brushes (for high grafting densities) is less well defined or not even reached. In this case, the applied load on the cantilever during TM-AFM height imaging of the thickness evaluation was too high for these soft brushes and already reduced the "real" hydrated thickness of polymer brushes. For PAAm brushes with a high degree of cross-linking the brushes may have become dehydrated resulting in such substantial stiffening that the soft AFM cantilevers used could not indent the brushes completely. This means, that by the use of conventional contact and tapping modes, the AFM tip significantly influences the thickness and lateral dimensions of polymer brushes.

Taking these assumptions into account the "real" hydrated thickness of PAAm brushes was measured as a depth difference between tip-brush contact point - when the underlying substrate is sensed upon further compression (as indicated by the pronounced increase of the slope of the applied load against penetration depth, since the AFM tip cannot indent the stiff glass substrate) from the load penetration depth profiles. In this case, the unperturbed configuration of the polymer brush chains can be reconstructed from the applied load-penetration depth curves at near zero penetration. For measurements of this type, the force curves with varied indentation thresholds (60-300 nm) were measured and analyzed, as shown in Figure 5.61.

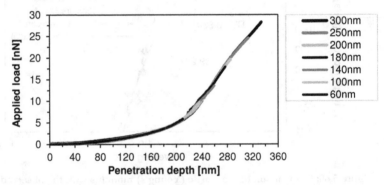

Figure 5.61: Applied load versus penetration depth curves for PAAm brushes with varied dimensions of the indentation threshold

The determination of the zero-separation distance is generally not possible, since the graft polymer layers are still compressed at the highest loads applied during the experiment. Hence there is always a compressed polymer layer between the tip and the sample substrate at the highest loads, the thickness of which cannot be measured independently and a certain thickness of the compressed brushes (the dry thickness of brushes) has to be added to the result. As expected, the hydrated thickness of PAAm brushes gained from TM-AFM height images (42 nm) was extremely lower than the zero separation distance derived from force-penetration depth analysis. Because in the curves of Figure 5.61 no vertical increase of the applied load can be recognized, even for the force curves with a threshold of 300 nm, the substrate was not reached. In this case polymer brushes containing only PAAm were not deep enough indented, to be analyzed in this way. When the zero-penetration distance of the other brushes is estimated out of Figure 5.56, new and higher swelling ratios are evaluated and collected in Table 5.5. But the substrate of polymer brushes with low fractions of cross-linker in the feed was not reached during the nanoindentations, whereas the results which are potentially underestimates should be handled with care. Accordingly, the new determined swelling ratios are still lower compared to the values reported by the group of Spencer et al. [91]

Table 5.5: Comparison of the new determined zero penetration distances and corresponding swelling ratios with the wet thicknesses determined from height images measured in TM-AFM

bisAAm	dry thickness [nm]		wet thickness [nm]		swelling ratio$_1$		zero penetration distance [nm]	swelling ratio$_2$
	average	S	average	S	average	S		
0	31.33	1.69	42.74	1.14	1.37	0.06	>250	7.98
0.015	16.95	0.57	22.49	0.85	1.33	0.02	-	-
0.025	14.98	0.71	18.77	0.75	1.25	0.02	>55	3.67
0.05	12.05	0.28	13.84	0.53	1.15	0.04	37	3.07
0.5	8.13	0.74	8.43	0.74	1.04	0.01	10	1.23
1	5.80	0.42	5.83	0.77	1.00	0.02	6	1.03

When the hydrated ODT monolayers of patterned PAAm brush samples were analyzed by AFM indentations, the force curves revealed no attractive interactions (snap in onto the surface) between the ODT monolayer and the tip, which were expected due to the hydrophobic nature of ODT. The corresponding deflection-piezo position curves are shown in Figure 5.62.

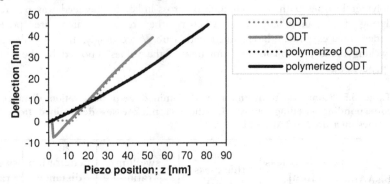

Figure 5.62: Raw nanoindentation curves of ODT monolayer with and without possible PAAm brushes on top

The raw force curves additionally show a non-linear behavior for the ODT monolayer, which was indented on a polymerized sample. In the beginning of this chapter FTIR measurements already revealed an existing PAAm layer on top of the monolayer, which should only constitute out of a dummy initiator. When the elastic modulus of this force curve is calculated, it corresponds to the elastic modulus of PAAm brushes, which has a dry thickness of 10 nm and is consistent with the evaluated thickness evaluated by FTIR. It should be taken into account that this persistent PAAm layers also have an impact on the thickness measurements of hydrated PAAm brushes by TM-AFM.

6 Discussion, Conclusions and Outlook

The aim of this thesis was the synthesis of PAAm brushes with varied polymerization time, grafting density and cross-linking density. Hence, the polymerizations of pure, mixed and patterned initiator SAMs on gold surfaces were carried out in an aqueous AAm solution (with a cross-linker in the feed) at ambient temperature (~23°C) via SI-ATRP. These polymer brushes were investigated using ellipsometry, contact angle measurements and FTIR spectroscopy. Later on the surface morphology, thickness and wettability was determined by AFM measurements. The nanoindentation measurements revealed the mechanical properties of polymer brushes by the determination of the elastic modulus out of force curves.

In this study the surface composition of polymer brushes had to be investigated before further experiments with the AFM were carried out. Therefore, the first part of this study was focused on the investigation of the thicknesses and compositions of the polymer brushes with varied polymerization time, grafting and cross-linking density.

The variation of the polymerization time during the polymer brush synthesis revealed a dramatically increase in brush thickness in early stages of the reaction. After a polymerization time of 60 min a plateau, around 50 nm brush thickness, was reached, whereupon the polymer brush thickness no further increases. The results are in line with the ones recently observed by Liu et al. They also grafted PAAm brushes on gold surfaces via SI-ATRP and analyzed the protein resistance of PAAm brushes with different thicknesses.[6] Brush thickness is just one parameter, which plays an important role in the application of polymer brushes as antifouling materials. Also, additional parameters have a great impact on the antifouling properties of these brushes, like the polymer brush morphology, which is influenced by the variation of grafting and cross-linking density. This concept has, so far to our knowledge, not yet been investigated with grafted PAAm brushes on gold via SI-ATRP.

The grafting density of grafted polymer brushes on surfaces was conveniently adjusted by varying the fraction of the initiator on the surface. In this case the gold substrates were immersed overnight in solutions out of two different thiols, which lead to the formation of the mixed SAMs with the preferential adsorption of the thiol with the dummy functionalized head group. The compositions of the mixed SAMs were studied with the help of the contact angle measurements, evaluated by the Cassie equation and Israelachvili equation. Both equations, with the focus on heterogeneous surfaces (Cassie) or homogeneous surfaces (Israelachvili), led to similar values of the composition and were consistent with the results of a previous project.[84]

Afterwards, the mixed SAMs on gold were used for the polymerization of AAm for 60 min at ambient temperature (~23 °C). It was observed that the thickness increases with the initiator density. Assuming that the variables (temperature, time and monomer concentration) are constant, the change of the ellipsometric thickness for increasing values of initiator on the surface is largely due to the increase in brush density. Therefore, the polymer brush morphology can be controlled by the initiator density. It is assumed that the polymer chains undergo a mushroom-to brush conformational transition above a critical surface density and that the transition forces the active ends of the polymer chains to stretch up and away from the surface. This result can be compared with the work of Genzer et al, who prepared different initiator densities by gradients generated with CMPE on silicon substrates and polymerized AAm by ATRP (Figure 1.1). He obtained the same trend of the polymer brush thicknesses measured by ellipsometry for PAAm brushes.[21] Due to the lack of a possibility to measure the molecular weight of grafted brushes directly on the substrate, it is difficult to discern whether the polymerization kinetics or the length of the PAAm chains, also depends on the grafting density. It was shown, that polymer brushes with low initiator density on the surface exhibited larger brush thickness for a longer polymerization time. In conclusion, the surface concentration of initiator molecules was systematically varied to allow the control of the grafting density of PAAm polymer brushes.

The cross-linking density of polymer brushes was adjusted in a suitable manner by varying the amount of bisAAm during the polymerization. The addition of the cross-linker (bisAAm) decreased the polymer brush thickness and altered the composition of thin polymer brushes dramatically. This was confirmed by the additional appearance of a peak at 1502 cm^{-1} in the FTIR spectra, which is characteristic for the cross-linker. In contrast, the growth rate was found to be contrary to the trend observed by Spencer et al.[12]

The second part of the study included first of all measurements by AFM, which gave insight about the surface morphology, thickness and wettability of polymer brushes polymerized on patterned substrates with varied polymerization time, grafting and cross-linking density. The measurements of the surface morphology of polymer brushes with varied grafting and cross-linking density revealed a decreasing surface roughness with increasing fraction of initiator on the surface or decreasing fraction of cross-linker during polymerization. In addition to the measurement of the dry brush thickness, wettability experiments were additionally performed as a function of the PAAm brush grafting and cross-linking density on the substrate. The aim was to provide more insight into the polymer packing on the surface. Polymer brushes with minimized grafting density exhibited a greater swelling ratio than the ones with high grafting density and the irreversible cross-linking of polymer brushes effectively restricted the swelling response. In contrast, the utilization of patterned ODT monolayer revealed several challenges, due to the deposition in the initiator solution prior the polymerization. Because of an exchange of the persistent ODT molecules with initiator molecules, PAAm layers were either chemisorbed or physisorbed onto the ODT monolayer and prevented an easy thickness determination of polymer brushes.

In the end nanoindentation measurements were performed with the AFM and used to investigate the mechanical properties of polymer brushes as also the real hydrated polymer brush thickness. It was found that the mechanical properties of PAAm brushes are greatly dependent on the initiator density on the surface as well as on the cross-linking density. With increasing amount of cross-linker in the feed or even a less dense initiator monolayer on the surface, the stiffness of PAAM brushes increases.[12,91] These results should be interpreted with care in the case of a possible substrate effect due to very thin polymer brushes.[96] The indentation measurements by AFM additionally revealed that pure hydrated PAAm brushes with a high grafting density have a thinner hydrated thickness than measured with AFM imaging.

In summary, it has been shown that for these polymer brushes of PAAm the film thicknesses, surface morphologies, wettabilities and mechanical properties vary with initiator and cross-linking density, prepared by SI-ATRP of mixed SAMs of two alkanethiols, respectively with and one without a terminal ATRP initiator group and varied fractions of cross-linker during polymerization.

Future work should address the extension of the latter methodologies, polymerization time, grafting and cross-linking density to prepare complex combinatorial brush structures with variable morphologies and mechanical properties. The determination of the hydrated polymer brush thickness by ellipsometry with the substrates placed into a solution cell would be an approach without penetrating the swollen brushes with an applied load as compared with AFM measurements. Another interesting but very labor intense aspect would be the estimation of the molecular weight of the polymer grafted to the surface. Either a solution polymerization of the polymer can be carried out or the polymer brush chains can be detached due to oxidation. Then they have to be analyzed by GPC.

The results outlined in this report have practical implications for the study of cellular interactions on PAAm brushes because cell-substrate interactions are known to influence various cell characteristics, such as migration and adhesion. The influence of the substrate on these characteristics can be largely attributed besides its chemical composition to its mechanical properties. But it is crucial to tune and quantify precisely the mechanical properties of those substrates. In this case, polymer brushes of PAAm are useful for the design of variable substrates considering their reproducible control during preparation. Patterned PAAm brushes are additionally an interesting method to advance the understanding of how cells interact with substrates. When the pattern are of the same length scale as individual cells, such that cells were seeded onto the substrates they preferentially bound to the printed islands and spread to conform to their size and shape. It has to be taken into account, that cell area not only regulates the proliferation, but also programmed cell death, known as apoptosis, as stated by Chen et al.[97]

7 References

(1) Luo, N.; Hutchison, J. B.; Anseth, K. S.; Bowman, C. N. *Macromolecules.* **2002**, *35 (7)*, 2487–2493.
(2) Boer, B. de; Simon, H. K.; Werts, M. P. L.; van der Vegte, E. W.; Hadziioannou, G. *Macromolecules.* **2000**, *33 (2)*, 349–356.
(3) Ingall, M. D. K.; Honeyman, C. H.; Mercure, J. V.; Bianconi, P. A.; Kunz, R. R. *J. Am. Chem. Soc.* **1999**, *121 (15)*, 3607–3613.
(4) Fundeanu, I.; van der Mei, H. C.; Schouten, A. J.; Busscher, H. J. *Colloids Surf., B.* **2008**, *64 (2)*, 297–301.
(5) Brown, A. A.; Khan, N. S.; Steinbock, L.; Huck, W. T. S. *Eur. Polym. J.* **2005**, *41 (8)*, 1757–1765.
(6) Liu, Q.; Singh, A.; Lalani, R.; Liu, L. *Biomacromolecules.* **2012**, *13 (4)*, 1086–1092.
(7) Coad, B. R.; Lu, Y.; Glattauer, V.; Meagher, L. *ACS Appl. Mater. Interfaces.* **2012**, *4 (5)*, 2811–2823.
(8) Milner, S. T. *Science.* **1991**, *251*, 905–914.
(9) Prucker, O.; Rühe, J. *Langmuir.* **1998**, *14 (24)*, 6893–6898.
(10) Ulman, A. *Chem. Rev.* **1996**, *96*, 1533–1554.
(11) Otsu, T. *J. Polym. Sci., Part A: Polym. Chem.* **2000**, *38*, 2121–2136.
(12) Li, A.; Benetti, E. M.; Tranchida, D.; Clasohm, J. N.; Schoenherr, H.; Spencer, N. D. *Macromolecules.* **2011**, *44 (13)*, 5344–5351.
(13) Kannurpatti, A. R.; Lu, S.; Bunker, G. M.; Bowman, C. N. *Macromolecules.* **1996**, *29*, 7310–7315.
(14) Ma, H.; Li, D.; Sheng, X.; Zhao, B.; Chilkoti, A. *Langmuir.* **2006**, *22 (8)*, 3751–3756.
(15) Nakayama, Y.; Matsuda, T. *Macromolecules.* **1996**, *29*, 8622–8630.
(16) Luo, N.; Metters, A. T.; Hutchison, J. B.; Bowman, C. N.; Anseth, K. S. *Macromolecules.* **2003**, *36 (18)*, 6739–6745.
(17) Schmelmer, U.; Paul, A.; Küller, A.; Steenackers, M.; Ulman, A.; Grunze, M.; Gölzhäuser, A.; Jordan, R. *Small.* **2007**, *3 (3)*, 459–465.
(18) Ma, H.; Jinho Hyun; Philip Stiller; Chilkoti, A. *Adv. Mater.* **2004**, *16 (4)*, 338–341.
(19) Schmidt, R.; Zhao, T.; Green, J.-B.; Dyer, D. J. *Langmuir.* **2002**, *18 (4)*, 1281–1287.
(20) Laibinis, P. E.; Nuzzo, R. G.; Whitesides, G. M. *J. Phys. Chem. (Journal of Physical Chemistry).* **1992**, *96*, 5097–5105.
(21) Wu, T.; Efimenko, K.; Genzer, J. *J. Am. Chem. Soc.* **2002**, *124 (32)*, 9394–9395.
(22) Lee, B. S.; Lee, J. K.; Kim, W.-J.; Jung, Y. H.; Sim, S. J.; Lee, J.; Choi, I. S. *Biomacromolecules.* **2007**, *8 (2)*, 744–749.
(23) Feng, W.; Chen, R.; Brash, J. L.; Zhu, S. *Macromol. Rapid Commun.* **2005**, *26 (17)*, 1383–1388.
(24) Ayres, N. *Polym. Chem.* **2010**, *1 (6)*, 769–777.
(25) Edmondson, S.; Wilhelm T. S. Huck. *Adv. Mater.* **2004**, *16 (15)*, 1327–1331.
(26) Jordan, R., Ed. *Advances in Polymer Science*; Springer-Verlag: Berlin/Heidelberg, 2006.
(27) Edmondson, S.; Osborne, V. L.; Huck, W. T. S. *Chem. Soc. Rev.* **2004**, *33 (1)*, 14–22.
(28) Barbey, R.; Lavanant, L.; Paripovic, D.; Schüwer, N.; Sugnaux, C.; Tugulu, S.; Klok, H.-A. *Chem. Rev.* **2009**, *109 (11)*, 5437–5527.
(29) Huang, W.; Baker, G. L.; Bruening, M. L. *Angew. Chem. Int. Ed.* **2001**, *40 (8)*, 1510–1512.
(30) Steenackers, M.; Küller, A.; Stoycheva, S.; Grunze, M.; Jordan, R. *Langmuir.* **2009**, *25 (4)*, 2225–2231.
(31) Jiang, J.; Wang, X.; Lu, X.; Lu, Y. *Applied Surface Science.* **2008**, *255 (5)*, 1888–1893.
(32) Otsu, T.; Ogawa, T.; Yamamoto, T. *Macromolecules.* **1986**, *19*, 2087–2089.

(33)   Shah, R. R.; Merreceyes, D.; Husemann, M.; Rees, I.; Abbott, N. L.; Hawker, C. J.; Hedrick, J. L. *Macromolecules*. **2000**, *33 (2)*, 597–605.
(34)   Baum, M.; Brittain, W. J. *Macromolecules*. **2002**, *35 (3)*, 610–615.
(35)   Wu, T.; Efimenko, K.; Vlček, P.; Šubr, V.; Genzer, J. *Macromolecules*. **2003**, *36 (7)*, 2448–2453.
(36)   Kim, J.-B.; Huang, W.; Miller, M. D.; Baker, G. L.; Bruening, M. L. **2002**, 386–394.
(37)   Hyun, J.; Kowalewski, T.; Matyjaszewski, K. *Macromol. Rapid Commun.* **2003**, *24 (18)*, 1043–1059.
(38)   Hucknall, A.; Rangarajan, S.; Chilkoti, A. *Adv. Mater.* **2009**, *21*, 2441–2446.
(39)   Jones, D. M.; Huck, W. T. S. *Adv. Mater.* **2001**, *13*, 1256–1259.
(40)   Kong, X.; Kawai, T.; Abe, J.; Iyoda, T. *Macromolecules*. **2001**, *34 (6)*, 1837–1844.
(41)   Matyjaszewski, K.; Xia, J. *Chem. Rev.* **2001**, *101 (9)*, 2921–2990.
(42)   Matyjaszewski, K.; Davis, K.; Patten, T. E.; Wei, M. *Tetrahedron*. **1997**, *53*, 15321–15329.
(43)   Xiao, D.; Wirth, M. J. *Macromolecules*. **2002**, *35 (8)*, 2919–2925.
(44)   Husseman, M.; Malmström, E. E.; McNamara, M.; Mate, M.; Mecerreyes, D.; Benoit, D. G.; Hedrick, J. L.; Mansky, P.; Huang, E.; Russell, T. P.; Hawker, C. J. *Macromolecules*. **1999**, *32 (5)*, 1424–1431.
(45)   Zhou, F.; Huck, W. T. S. *Phys. Chem. Chem. Phys.* **2006**, *8 (33)*, 3815–3823.
(46)   Chechik, V.; Crooks, R. M.; Stirling, C. J. *Adv. Mater.* **2000**, *12*, 1161–1171.
(47)   Xia, Y.; Zhao, X.-M.; Whitesides, G. M. *Microelectronic Enigneering*. **1996**, *32*, 255–268.
(48)   Schreiber, F. *Prog. Surf. Sci.* **2000**, *65*, 151–256.
(49)   Bain, C. D.; Troughton, E.; Tao, Y.; Evall, J.; Whitesides, G. M.; Nuzzo, R. G. *J. Am. Chem. Soc.* **1989**, *111*, 321–335.
(50)   Zhou, F.; Zheng, Z.; Yu, B.; Liu, W.; Huck, W. T. S. *J. Am. Chem. Soc.* **2006**, *128 (50)*, 16253–16258.
(51)   Bain, C. D.; Evall, J.; Whitesides, G. M. *J. Am. Chem. Soc.* **1989**, *111*, 7155-7154.
(52)   Bain, C. D.; Whitesides, G. M. *J. Am. Chem. Soc.* **1988**, *110*, 6560–6561.
(53)   John P. Folkers; Laibinis, P. E.; Whitesides, G. M. *Langmuir*. **1992**, *8*, 133–1341.
(54)   Bayat, H.; Tranchida, D.; Song, B.; Walczyk, W.; Sperotto, E.; Schoenherr, H. *Langmuir*. **2011**, *27 (4)*, 1353–1358.
(55)   Ruiz, S. A.; Chen, C. S. *Soft Matter*. **2007**, *3 (2)*, 168–177.
(56)   Perl, A.; Reinhoudt, D. N.; Huskens, J. *Adv. Mater.* **2009**, *21 (22)*, 2257–2268.
(57)   Wilbur, J. L.; Kumar, A.; Biebuyck, H. A.; Enoch Kim; Whitesides, G. M. *Nanotechnology*. **1996**, *7*, 452–457.
(58)   Kumar, A.; Whitesides, G. M. *Appl. Phys. Lett.* **1993**, *63 (14)*, 2002–2004.
(59)   Xia, Y.; Tien, J.; Whitesides, G. M.; Quin, D. *Langmuir*. **1996**, *12*, 4033–4038.
(60)   Ma, H.; Wells, M.; Beebe, T. P.; Chilkoti, A. *Adv. Funct. Mater.* **2006**, *16 (5)*, 640–648.
(61)   Bass, R. B.; Lichtenberger, A. W. *Applied Surface Science*. **2004**, *226 (4)*, 335–340.
(62)   Cringus-Fundeanu, I.; Luijten, J.; van der Mei, H. C.; Busscher, H. J.; Schouten, A. J. *Langmuir*. **2007**, *23 (9)*, 5120–5126.
(63)   Bao, Z.; Bruening, M. L.; Baker, G. L. *Macromolecules*. **2006**, *39 (16)*, 5251–5258.
(64)   Jones, D. M.; Brown, A. A.; Huck, W. T. S. *Langmuir*. **2002**, *18 (4)*, 1265–1269.
(65)   Liu, H.; Li, M.; Lu, Z.-Y.; Zhang, Z.-G.; Sun, C.-C. *Macromolecules*. **2009**, *42 (7)*, 2863–2872.
(66)   Yamamoto, S.; Ejaz, M.; Tsujii, Y.; Fukuda, T. *Macromolecules*. **2000**, *33 (15)*, 5608–5612.
(67)   Loveless, D. M.; Abu-Lail, N. I.; Kaholek, M.; Zauscher, S.; Craig, S. L. *Angew. Chem. Int. Ed.* **2006**, *45 (46)*, 7812–7814.
(68)   Cassie, A. B. D. *Discuss. Faraday Soc.* **1948**, *3*, 11–16.
(69)   Cassie, A. B. D.; Baxter, S. *Trans. Faraday Soc.* **1944**, *40*, 546–551.
(70)   Drummond, C.; Israelachvili, J. *J. Pet. Sci. Eng.* **2002**, *33*, 123–133.
(71)   Wenzel, R. N. *Ind. Eng. Chem.* **1936**, *28*, 988–994.
(72)   Wenzel, R. N. *J. Phys. Colloid. Chem.* **1949**, *53 (9)*, 1466–1467.

(73) Butt, H.-J.; Cappella, B.; Kappl, M. *Surf. Sci. Rep.* **2005**, *59 (1-6)*, 1–152.
(74) Tranchida, D.; Kiflie, Z.; Piccarolo, S. *Formatex.* **2007**.
(75) Tranchida, D.; Piccarolo, S.; Loos, J.; Alexeev, A. *Macromolecules.* **2007**, *40 (4)*, 1259–1267.
(76) Schoenherr, H.; Ringsdorf, H. *Langmuir.* **1996**, *12*, 3891–3897.
(77) Hampton, J. R.; Dameron, A. A.; Weiss, P. S. *J. Am. Chem. Soc.* **2006**, *128 (5)*, 1648–1653.
(78) Song, B.; Walczyk, W.; Schoenherr, H. *Langmuir.* **2011**, *27 (13)*, 8223–8232.
(79) Stamou, D.; Gourdon, D.; Liley, M.; Burnham, N. A.; Kulik, A.; Vogel, H.; Duschl, C. *Langmuir.* **1997**, *13*, 2425–2428.
(80) Samorí, P.; Diebel, J.; Löwe, H.; Rabe, J. P. *Langmuir.* **1999**, *15 (7)*, 2592–2594.
(81) Hegner, M.; Wagner, P.; Semenza, G. *Surf. Sci.* **1993**, *291*, 39–46.
(82) Wagner, P.; Hegner, M.; Güntherodt, H.-J.; Semenza, G. *Langmuir.* **1995**, *11*, 3867–3875.
(83) Fundeanu, I.; Klee, D.; Schouten, A. J.; Busscher, H. J.; van der Mei, H. C. *Acta Biomaterialia.* **2010**, *6 (11)*, 4271–4276.
(84) Lilge, I. *Bachelor Thesis.* **2010**, Universitaet Siegen.
(85) Berchtold, K. A.; Lovell, L. C.; Nie, J.; Hacioglu, B.; Bowman, C. N. *Polym. J.* **2001**, *42*, 4925–4929.
(86) Cohen Stuart, M. A.; Vos, W. M. de; Leermakers, F. A. M. *Langmuir.* **2006**, *22 (4)*, 1722–1728.
(87) Gong, J. P.; Kurokawa, T.; Narita, T.; Kagata, G.; Osada, Y.; Nishimura, G.; Kinjo, M. *J. Am. Chem. Soc.* **2001**, *123 (23)*, 5582–5583.
(88) Gong, J. P. *Soft Matter.* **2006**, *2 (7)*, 544–552.
(89) Wang, X.; Tu, H.; Braun, P. V.; Bohn, P. W. *Langmuir.* **2006**, *22 (2)*, 817–823.
(90) Moh, L. C. H.; Losego, M. D.; Braun, P. V. *Langmuir.* **2011**, *27 (7)*, 3698–3702.
(91) Li, A.; Ramakrishnaa, S. N.; Kooij, E. S.; Espinosa-Marzala, R. M.; Spencer, N. D. *Soft Matter.* **2012**, *8*, 9092–9100.
(92) Huang, X.; Doneski, L. J.; Wirth, M. J. *Anal. Chem.* **1998**, *70*, 4023–4029.
(93) Gao, X.; Kučerka, N.; Nieh, M.-P.; Katsaras, J.; Zhu, S.; Brash, J. L.; Sheardown, H. *Langmuir.* **2009**, *25 (17)*, 10271–10278.
(94) Tranchida, D.; Kiflie, Z.; Piccarolo, S. *Formatex.* **2007**.
(95) Boudou, T.; Ohayon, J.; Picart, C.; Tracqui, P. *Biorheology.* **2006**, *43*, 721–728.
(96) Tranchida, D.; Lilge, I.; Schoenherr, H. *Polym Eng Sci.* **2011**, *51 (8)*, 1507–1512.
(97) Alom Ruiz, S.; Chen, C. S. *Soft Matter.* **2007**, *3 (2)*, 1–11.
(98) Heeb, R.; Bielecki, R. M.; Lee, S.; Spencer, N. D. *Macromolecules.* **2009**, *42 (22)*, 9124–9132.
(99) Harris, B. P.; Metters, A. T. *Macromolecules.* **2006**, *39 (8)*, 2764–2772.
(100) Rahane, S. B.; Kilbey, M. S.; Metters, A. T. *Macromolecules.* **2005**, *38 (20)*, 8202–8210.
(101) Rahane, S. B.; Metters, A. T. *Macromolecules.* **2006**, *39 (26)*, 8987–8991.

8 Appendix

8.1 PAAm Brushes via PMP

This chapter focuses on the first route of preparation of PAAm brushes via PMP on the basis of Spencer et al.[12] Beginning with the route of synthesis of the iniferter which encountered several problems. These problems were analyzed and tried to be solved.

8.1.1 Iniferter

The iniferter, N,N-(diethylamino)dithiocarbamoylbenzyl(trimethoxy)silane (SBDC), is a functionalized silane, which is used to perform the PMP and was synthesised from commercial starting materials 4-(Chloromethyl)phenyltrimethoxysilane [purchased from Alfa Aesar, 90% pure] and Diethyl-dithiocarbamicacid diethylammoniumsalt [purchased from Aldrich].

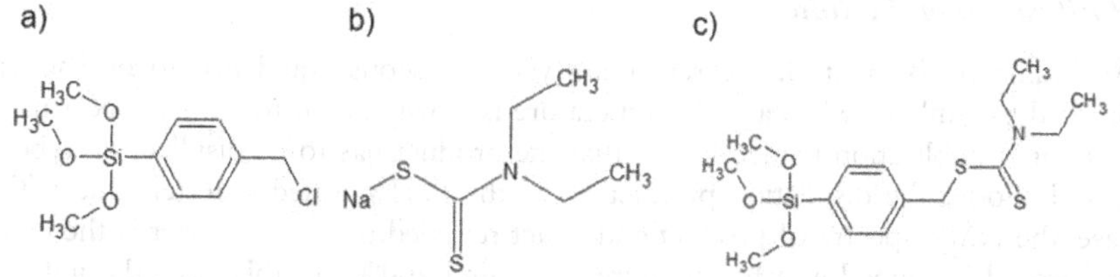

Figure 8.1: Molecular structure of a) 4-(Chloromethyl)phenyl-trimethoxysilane/ Diethyldithiocarbamicacid diethyl-ammoniumsalt, b) N,N-diethyldithiocarbamate (STC) and c) Molecular structure of N,N-(diethylamino) dithiocarbamoylbenzyl (trimethoxy)silane (SBDC)

Synthesis of the Iniferter

The photoiniferter was synthesized according to de Boer et al.[2] The silane (1.48 g, 6 mM) and the STC (1.02 g, 6 mM) were each dissolved separately in 10 mL of dry THF. All used glassware for synthesis was dried in an oven for an hour at 120 °C. After injection of the silane solution the three-necked reaction flask was sealed with rubber septen. A needle, attached to the argon gas line was added to the first septum, the second septum was kept free for filling and another needle was introduced through the third septum to prevent pressure build up due to evaporation of THF. The STC solution was added slowly via a syringe to the dissolved silane solution in the reaction flask while being stirred for 3 hours at room temperature. Almost immediately a white precipitate (NaCl) was formed, and the solution became more yellow.[2]

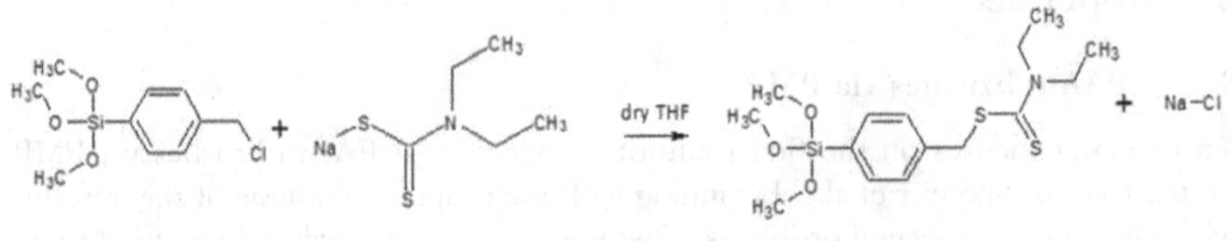

Figure 8.2: Synthesis of *N,N*-(diethylamino)dithiocarbamoylbenzyl(trimethoxy)silane (SBDC)

The precipitate was removed by passing the solution through a Teflon syringe filter (0.2 μm) under argon gas flow. To eliminate moisture which will react with and degrade the silane groups, the next synthesis of the iniferter was later on performed in the glove box. The THF was evaporated under reduced pressure by a rotary vane pump.[98]

Purification of the Iniferter

According de Boer et al. the remaining yellow viscous liquid had to be vacuum distilled (0.1 mbar), whereas the temperature is slowly raised to 220 °C.[2] Normally, when in a publication is mentioned, that the product has to be distilled, everybody would choose the distillation product to be the final purified iniferter. But in this case, the NMR spectra of product and educt revealed that the iniferter is the educt and would be ruined at a temperature of around 200°C. In this case, the iniferter was later on only purified by attaching the vial to a high vacuum line over night. The photoiniferter was obtained as a yellow viscous liquid (30%) and was characterized by ¹H NMR.

¹H-NMR
Fourier transformed ¹H NMR spectra were recorded in CDCl₃ with chloroform as internal reference at 7.24 ppm and the data were in accordance with the values obtained by de Boer et al.

Table 8.1: Important ¹H NMR values

[ppm]	pattern	number of H	signal assignment
7.65-7.38	dd	4H	C₆H₄
4.55	s	2H	CH₂S
4.05	q	2H	NCH₂
3.73	q	2H	NCH₂
3.62	s	9H	Si(OCH₃)₃
1.25	t	6H	CH₃

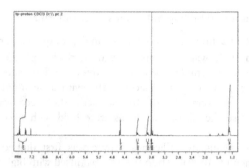

Figure 8.3: ¹H NMR spectra of SBDC

Ultraviolet – Visible (UV-VIS) Spectroscopy

10 μL of SBDC was dissolved in 1 mL of acetone and measured against an acetone reference. AAm was dissolved in water (1M) and measured against a water reference.

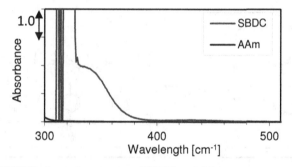

Figure 8.4: UV-VIS absorption spectra of SBDC and AAm

Figure 8.4 shows the UV-VIS absorption spectra of both the SBDC photoiniferter and acrylamide. The obtained spectra were taken as a reference for the selection of the spectral UV range in order to ensure effective initiation while avoiding polymerization of the monomer solution. As it can be seen, the monomer and the SBDC photoiniferter have distinct regions of UV absorption essential for a controlled surface initiated polymerization process. The UV source has minimal emission in the region where the monomer shows significant UV absorptions. Figure 8.4 indicates that UV induced polymerization for AAm in solution is unlikely to occur since the narrow emission spectrum of the UV-LED does not overlap with the UV absorption region of the monomer.

8.1.2   Modification of Silicon Substrates with Iniferter

To be able to obtain a homogeneous film of initiators on the surface, the silicon substrates [Silicon (111) wafers; Okmetric wafers, thickness 525 ± 25 µm] were cleaned before the self-assembling process. Therefore a silicon substrate, with a size of 1.5 x 1.5 cm, was cut out of a silicon wafer with a diamond glass cutter. In addition, to avoid any scratches on the surfaces, the piece was handled very carefully. In this case the silicon substrates were hold with tweezers only at the edges.

In the beginning, the silicon substrate was first dipped and washed in chloroform and then in ethanol. Afterwards it was rinsed with high purity water and then dipped for 1 min into the Piranha-Solution (1:3 (v/v) 30% H_2O_2 and concentrated H_2SO_4), to introduce reactive hydroxyl functionalities on the silicon surface. After a second rinsing with high purity water and with ethanol, the silicon substrate was dried with a clean and dry nitrogen gas flow.[99,100] Then the clean silicon substrate was immersed in 2ml iniferter solution (5 mM) in a oven dried snap-on lid glass.[12]

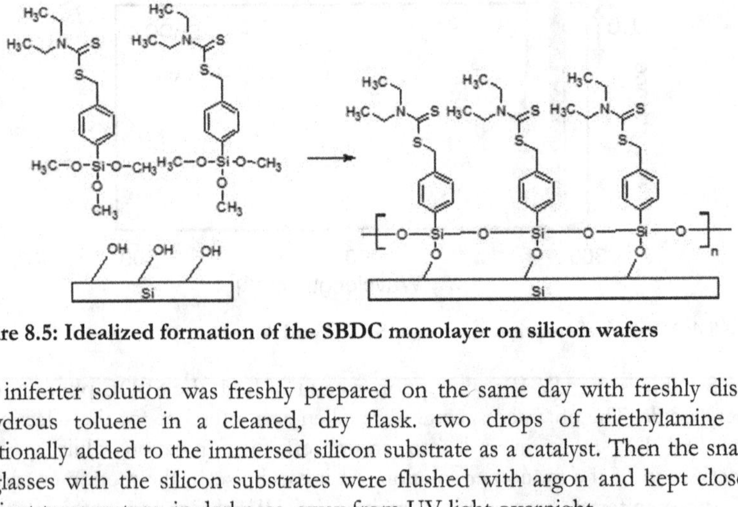

Figure 8.5: Idealized formation of the SBDC monolayer on silicon wafers

The iniferter solution was freshly prepared on the same day with freshly distilled anhydrous toluene in a cleaned, dry flask. two drops of triethylamine were additionally added to the immersed silicon substrate as a catalyst. Then the snap-on lid glasses with the silicon substrates were flushed with argon and kept closed at ambient temperature, in darkness, away from UV light overnight.

The SAMs were thoroughly rinsed the other day with copious amounts of pure toluene and afterwards sonicated in pure toluene for 40 min to remove unreacted SBDC and reaction byproducts. Later on the substrates were again rinsed with pure toluene and dried in a stream of nitrogen. Finally the substrates were kept in a petri dish and stored at RT in a dark place until polymer grafting.

The formation of a uniform SBDC monolayer on silicon wafers was investigated by measuring the contact angles of the cleaned and modified substrates. Silicon wafers have typical water contact angles smaller than 10° (measured: 5 ± 3°), when the Si surface is cleaned with organic solvents. For the SBDC-modified substrates a significant higher contact angles were obtained, which can be related to a considerable hydrophobic effect of SBDC. The values of the contact angles of SBDC SAMs collected from different publications are summarized in the following table.

Table 8.2: Contact angle collection of SBDC modified silicon substrates

origin	contact angle
de Boer et al.[2]	80 ± 5°
Li et al.[12]	70 ± 2°
Heeb et al.[98]	68 ± 3°
Harris et al.[99]	65 ± 3.2°
Rahane et al.[101]	64 ± 3°

The obtained water contact angles are in good agreement with the previously reported values that were obtained by 73 ± 4°, which should be handled with care since the values represented in Table 8.2 open a possible range between 61° and 85°. The presence of an organic thin layer was additionally confirmed by ellipsometry. An average ellipsometric thickness of 1.1 ± 0.1 nm was measured.

Table 8.3: Thicknesses of the SBDC monolayer characterized by ellipsometry

origin	ellipsometric thickness
de Boer et al.[2]	~1 nm
Li et al.[12]	1.1 ± 0.1 nm
Heeb et al.[98]	0.75 ± 0.06 nm
Harris et al.[99]	1.5 ± 0.4 nm
Rahane et al.[100]	1.4 ± 0.2 nm

The thickness values are consistent with the previously observed values and values predicted by bond length calculations (1.3 nm). In the case of a lower value than the theoretical maximum thickness, calculated by Rahane et al., suggests that the surface coverage is below that of a full monolayer. In contrast, a too high concentration of surface radicals can lead to extensive termination reactions in surface initiated polymerization approaches.[100]

8.1.3 Construction of the UV-LED setup

For the polymerization of the SBDC modified substrates in an aqueous solution und the irradiation of an UV-LED a suitable setup had to be constructed. A UV-LED was chosen due to its sharp emission spectrum and ease in selecting a specific wavelength region according to the employed photo-initiating system and the reactive monomer. AAs well as the necessity for optical filters to block irradiation of undesired regions in the lamp spectrum is greatly reduced, because the use of such filters frequently reduced the intensity, which leads to very poor energy efficiency and long irradiation times. The low wavelength region of unfiltered mercury lamps generally induces polymerization of the monomer in bulk solution, which can be excluded by the application of an UV-LED. In our case an additional cooling system was attached to the UV-LED to avoid a warming up by the LED (Figure 8.6).

Flasks have in general no flat glass window on the top, whereas a special Teflon reaction cell was build. The reaction cell had to have gas inlet, to be able to work in an argon atmosphere, as well as a fluid inlet, to inject the reaction medium (indicated with a red and blue arrow in Figure 8.6a).

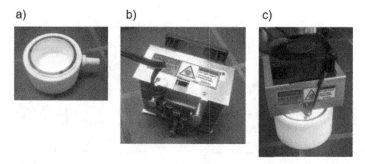

Figure 8.6: Images of a) the reaction cell, b) the UV-LED with cooling system and c) the whole reaction set up for a PMP

8.1.4   Preparation of Polymer Brushes

For the synthesis of polymer brushes, an aqueous solution (1 M) of the acrylamide (AAm) was prepared in an oven dried glass Schlenk tube and sealed with a rubber septum. The monomer solution was degassed by subjecting it to three freeze-vacuum-thaw cycles. Thereby the solution is frozen by immersion of the flask in liquid N_2. When the solvent is completely frozen, the flask is opened to the vacuum and pumped for 2-3 min. The flask is then closed and warmed until the solvent has completely melted. This procedure is repeated for three times and after the last cycle the flask is backfilled with argon gas.

Upon polymerization, a SBDC modified substrate was placed in a reaction cell made of Teflon. The volume of the cell is 20 ml. The cell was covered with a borosilicate glass plate and sealed using a seated O-ring and parafilm. A long steel needle attached to a 10 ml glass syringe was introduced through the septum into the Schlenk tube to aspirate out uniformly monomer solution. Then the monomer solution was transferred through a hole to the reaction cell containing the SBDC modified substrate.[12]

The polymerization was started immediately by the irradiation of a self-made UV-LED setup. The setup incorporated a LED with a narrow emission spectrum at 365 ± 5 nm, which was cooled by a fan to maintain the temperature at ambient conditions. An irradiation distance of 3 cm resulted from the developed reaction vessel. To prevent oxygen from entering the system the reaction chamber was continuously purged with argon gas.

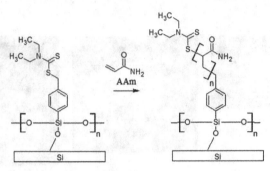

Figure 8.7: Idealized polymerization with AAm

After the photopolymerization, with duration of 30 min, the polymerized substrates were rinsed with copious amounts of water to stop the polymerization. Then the wafers were rinsed with pure ethanol and blown dry with a clean and dry nitrogen gas flow.

The formation of PAAm brushes on silicon wafers was investigated by contact angle measurements, whereas the polymer brush surface exhibited a contact angle of 31.5 ± 0.8°. The thickness of PAAm brushes was determined by ellipsometry and revealed 5.4 nm. This result is far too thin for PAAm brushes polymerized for 30 min, which should have a thickness around 400 nm according to Spencer et al.[12] Solution polymerizations with the iniferter, as well as SI-ATRP with AAm confirmed that the setup is in working order, as well as the polymerizability of the monomer and the N,N-diethyldithiocarbamate.

8.2   PAAm Brushes via ATRP

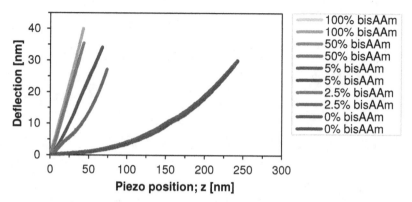

Figure 8.8: Raw nanoindentation curves of PAAm brushes with varied cross-linker (bisAAm) in the feed during polymerization

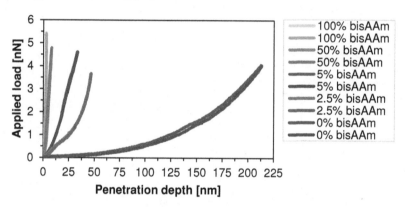

Figure 8.9: Applied load versus penetration depth curves for PAAm brushes with varied cross-linker (bisAAm) in the feed during polymerization

Printed in the United States
By Bookmasters